Nanostructuring Operations in Nanoscale Science and Engineering

Kal Renganathan Sharma, PE

New York   Chicago   San Francisco
Lisbon   London   Madrid   Mexico City
Milan   New Delhi   San Juan
Seoul   Singapore   Sydney   Toronto

The *McGraw·Hill* Companies

Cataloging-in-Publication Data is on file with the Library of Congress.

McGraw-Hill books are available at special quantity discounts to use as pre-miums and sales promotions, or for use in corporate training programs. To contact a representative please e-mail us at bulksales@mcgraw-hill.com.

Nanostructuring Operations in Nanoscale Science and Engineering

1 2 3 4 5 6 7 8 9 0   DOC/DOC   0 1 4 3 2 1 0 9

ISBN 978- 0-07-162295-0
MHID 0-07-162295-0

The pages within this book were printed on acid-free paper.

Sponsoring Editor	**Copy Editor**	**Composition**
Taisuke Soda	Susan Fox-Greenberg	International Typesetting and Composition
Acquisitions Coordinator	**Proofreader**	
Michael Mulcahy	Bhavna Gupta	**Art Director, Cover**
Editorial Supervisor	**Indexer**	Jeff Weeks
David E. Fogarty	Robert Swanson	
Project Manager	**Production Supervisor**	
Vipra Fauzdar	Richard C. Ruzycka	

The book is dedicated to
R. Hari Subrahmanyan Sharma (alias Ramkishan),
my eldest son,
who turns eight on August 13, 2009,
with unconditional love.

About the Author

Dr. Kal Renganathan Sharma, PE holds joint appointments as adjunct professor in three departments, Chemical Engineering, Mechanical Engineering, and Civil and Environmental Engineering, at Prairie View A&M University, Prairie View, Texas. He is the author of 6 books, 13 journal articles, 454 conference papers, 53 preprints, 36 seminars/invited lectures, and 21 other publications. He has instructed more than 2000 students in India and the United States via 70 semester courses. He received all of his three degrees in chemical engineering—B.Tech. from Indian Institute of Technology, Chennai, India in 1985 and MS and PhD from West Virginia University, Morgantown, West Virginia in 1987 and 1990, respectively. He has held a number of high-level positions in engineering colleges and universities. He has served as a reviewer for John Wiley and Sons, New York, *Journal of Thermophysics and Heat Transfer, Chemical Engineering,* and *Chemical Engineering Communications.* He is a co-convener at the II International Workshop in Nanotechnology and Health Care, SASTRA University, Thanjavur, India and co-chair at a number of conferences including Track Co-Chair at the Energy Sustainability/Summer Heat Transfer Conference at San Francisco, California, in July 2009. Civic activities include President, India Students Association at West Virginia University, Morgantown, West Virginia, and General Secretary, Ganga, Indian Institute of Technology, Chennai, India.

Contents

Foreword

Predictions indicate that over one million scientists, engineers, technologists, and associated professional people will work in the nanoscience and nanotechnology related industries over the next 10 to 15 years, and the market for nanotechnology products will reach some $3 trillion by the year 2015. Engineers trained in nanotechnology will attract higher salaries compared with even computer hardware and chemical engineers. Indeed, nanoscale science and technology promise major advances in almost every area of our socioeconomic environment.

For instance, we can expect the following: the development of hybrid electric vehicles and highly energy efficient transport systems that can lower the cost of commuter traffic; new photovoltaics that can collect solar power efficiently and cost effectively in deserts regions; the creation of materials with thermal conductivity much higher than copper and aluminum; the synthesis of supremely tough nanocomposite materials; a paradigm shift in the effectiveness of our medical strategies; major advances in the power of nanomagnetic materials; and the development of molecular computers.

The microprocessor and personal computer revolution was largely a consequence of our success in miniaturization and until now computing speed has doubled every 18 months as more and more transistor elements have been packed on a silicon chip. Genetic technology is now showing analogous advances as the efficiency with which we are able to handle DNA increases dramatically. At the nanoscale, quantum mechanical effects occur that, if they can be harnessed effectively, promise novel and powerful new applications that are quite different from those with which we are familiar at the macroscopic level.

The all-carbon hollow cage molecules, the fullerenes, and their elongated cousins the carbon nanotubes (CNT) are stable allotropes that, in addition to graphene, graphite, and diamond, show fascinating promise as basic materials for novel nanoscale applications. The morphology of materials is a fascinating field and structure-related properties are of key interest in product development and process engineering resulting in materials with advanced performance in sustainable, environmentally friendly applications.

If all these exciting advances are to be realized, then the next cohort of young scientists, engineers, and technologists must have a sound education in nanoscale science and technology and this education needs to be integrated into undergraduate curricula in chemical engineering. This text is a welcome and highly effective response to this challenge that must be met; if we are to develop the sustainable, technologies we shall certainly need to survive into the next century.

Harold Kroto
Department of Chemistry and Biochemistry
Florida State University
Tallahassee, Florida
kroto@chem.fsu.edu
www.kroto.info

Preface

I served as a co-convener along with Dr. S. Swaminathan of The II International Workshop on Nanotechnology and Health Care, conducted at SASTRA University, Thanjavur, India in May 2005. The President of India Dr. A. P. J. Abdul Kalam dedicated the Center for Nanotechnology and Advanced Biomaterials to the Indian Nation in September 2006. I taught Introduction to Nanotechnology and Nanofabrication Techniques to graduate students in 2005–2006 and 2006–2007 in India.

In the summer of 2008, Dr. Irvin Osborne Lee, Head of the Department of Chemical Engineering at Prairie View A&M University, Prairie View, Texas, charted me with the task of integrating advances in nanotechnology into the chemical engineering curriculum with a project on interlinked curriculum component (ICC) on nanotechnology. Three campuses were collaborating—Texas A&M University, College Station, Texas, Texas A&M at Kingsville, and Prairie View A&M University at Prairie View, Texas. It was clear that a good textbook in the area of nanostructuring operations was not available. This textbook is designed to cater to the students of nanotechnology the world over and the practitioners in the industry that work in nanostructuring operations. A chapter on nanoscale effects in time domain in heat conduction is also included. Review questions are provided at the end of each chapter, totaling approximately 550.

Acknowledgments

Such a project would not have been possible without years of training and financial support from a number of resources. I would like to start with my high school science and math teachers who taught me to think independently. During my B. Tech. at Indian Institute of Technology, Chennai, India, eminent scholars such as M. S. Ananth (currently the Director at IIT-Chennai), Prof. Y. B. G. Varma, Dr. Durga Prasad Rao, Dr. Neelakantan, Dr. Venkatram, Dr. Krishnaiah, Prof. C. A. Sastry, Prof. Ramachandra Rao, and Dr. Baradarajan, among others, trained me well and new courses as the field emerged were nothing new to us. As a graduate student at West Virginia University, Morgantown, West Virginia, I had the privilege of studying under Prof. Dady B. Dadyburjor (currently Chairman), Prof. E. V. Cilento (currently, Dean), Prof. W. B. Whiting (student of J Prausnitz), Prof. J. A. Shaeiwitz, Prof. R. Yang (student of Lapidus), Prof. John W. Zondlo, and Prof. R. Turton (student of O. Levenspiel). John W. Zondlo funded my first conference paper presentation at Maastricht, Netherlands. As a PhD student, I presented a few conference papers for R. Turton, who funded my flight to San Francisco, California from Morgantown, West Virginia.

I worked in Monsanto Plastics Technology under Dr. Victoria Franchetti Haynes at Indian Orchad, Massachusetts. She was highly supportive of ventures such as nanostructuring operations. She funded my travel to AIChE conferences in Chicago, Illinois, in 1990, Los Angeles, California, in 1991, and Miami, Florida, in 1992 and to present papers at the technology symposiums at St. Louis, Missouri. In October 1995, I was on a transatlantic flight funded by R. Shankar Subramanian, chairman of chemical engineering at Clarkson University, Potsdam, New York, as his postdoctoral research associate for BDPU test validation in Officine Galileo, Turin, Italy, and tracer particle development in Frieberg, Germany.

Nason Pritchard funds were granted for my paper presentation at the World Congress in chemical engineering in San Diego, California, in 1996 as Adjunct Assistant Professor from West Virginia University. In 1998 and 1990, I presented 90 conference papers at 19 major conferences. Special thanks to Edward J. Wegmann, Chairman, Advanced Engineering Statistics, George Mason University and family and friends, not all of whom can be mentioned here due to space limitations. Special mention of Prof. Nithi T. Sivaneri, Associate Chair, Mechanical & Aerospace Engineering Department, West Virginia University, for his support.

During my instruction in India at Vellore Institute of Technology, Vellore, India, as Head of Department, and at SASTRA University, Thanjavur, India, as professor, Hon. G. Viswanathan, former minister in the government of Tamil Nadu, and Sri. R. Sethuraman, former Trustee of Madam of Sankaracharyas of Kanchi, were supportive. My travel to present papers at national conferences every year and in New Orleans, Louisiana, in 2003 and Atlanta, Georgia, in 2006 were funded by endowments.

Special thanks again to Dr. Irvin Osborne-Lee for funding my travel to present papers at the 99th AIChE Annual Meeting in Salt Lake City, Utah, in November 2007; colocated ACS and AIChE meetings in New Orleans in April 2008; Summer Heat Transfer Conference and Nanotechnology Conference at Jacksonville, Florida, in August 2008; and Southwest Regional Meeting of the American Chemical Society, Little Rock, Arkansas, in October 2008.

Kal Renganathan Sharma, PE

CHAPTER 1

Introduction

Learning Objectives

- Scope of nanoscale science and engineering
- Raleigh criterion for resolution
- Commercial products
- Feynman's vision on miniaturization and data storage
- Drexler-Smalley debate on molecular assemblers
- Salient events in the emergence of nanotechnology
- Applications
- Limits of miniaturization
- Thermodynamic stability of nanostructures
- Challenges of characterization of nanostructures

Nanoscale science and engineering pertains to the synthesis, characterization, and application of matter with at least one or more dimensions less than 100 nm. Materials made with dimensions on the nanoscale offer unique and different properties compared to those in the macroscale with conventional technology. The field holds a lot of promise not only for the practitioner of the technology but also for inventors and scientists. The emergence of nanotechnology attempts to tap into the benefits of miniaturization as realized in the microprocessor and microelectronics industry as part of the computer revolution. Research centers in nanotechnology are established with funding from the government in several countries around the world such as the United States, Western Europe, Japan, India, China, Brazil, and Russia. Computers that rival the brain in communications and storage of information, molecular motors, cellular machines, and drugs that target specific cells are expected to be developed with the growth of nanotechnology. The products shall span the power, biotechnology, computing, electronic, photonic, and manufacturing industries. It was conventional wisdom that the minimum size achievable was in the order of the wavelength of light. According to

1

Raleigh criterion, the minimum resolution size achievable was half the wavelength of light, i.e., 200 nm. The Prescott introduced by Intel in the Pentium IV chip has the dimensions of approximately 90 nm, lower than the previous minimum size of 130 nm. When integrated circuits (ICs) are printed with the dimensions of the images smaller than the wavelength of light, the Raleigh criterion is reinterpreted. Recent technological advances have affected a paradigm shift in the box thinking of the past.

Nanotechnology is derived from the French words *nanos*, which means dwarf, and *technologia*, which means a systematic treatment of an art of craft. A micron (or micrometer) is 1000 nm. A nanometer is one billionth of a meter (10^{-9} m). Comparing the dimensions of some common objects with some nanoscale objects can put things in perspective. The carbon nanotube (CNT) is 1 nm in diameter, a DNA molecule is 2 to 3 nm in diameter, and a house is 10 m wide. A raindrop is 2 to 3 mm in diameter, bacterium is 2 μm in length, and a strand of hair is 100 μm in diameter.

Advances in nanoelectronics, nanophotonics, and nanomagentics have seen the arrival of nanotechnology as a distinct discipline in its own right. As so many times in the past, clear vision preceded the invention. For example, *20,000 Leagues under the Sea,* a work of fiction, was published before the invention of the submarine. The television show *Star Trek* preceded space voyages. The 1966 film, *Fantastic Voyage,* where a group of doctors are shrunk to microscopic size and enter the body of a patient in a submarine-like capsule to set him right from the inside came well before nanotechnology. Nanoporous catalysts that provide increased surface area have been used. Molecular sieves with pore dimensions of the nanometer can be used to desalinate seawater when tailor-made to sift between the molecular sizes of water and common salt. A high school student returned home after his first lesson on Avogadro's number. Taking iron, for example, he found that 1 attogram of iron contained 602 molecules to fit into a cube with a side of 1 nm. Preparation at such dimensions is of increased interest to nanotechnologists. Nanostructure can be defined as an object with at least two of its dimensions less than 100 nm. The size range of atoms is between 1 and 4 A. Biomacromolecules such as DNA have a diameter of 2 nm. Proteins are 1 to 20 nm in size.

1.1 Commercial Products

The first *clay* nanocomposite was commercialized by Toyota. A Nylon-6 matrix filled with 5 percent clay is used for heat resistant timing belt covers. The material possesses 40 percent higher tensile strength, 60 percent higher flexural strength, 12.6 percent higher flexural modulus, and higher heat distortion temperature ranging from 65 to 152°C. General Motors (GM) commercialized the first exterior trim application of nanocomposite in their 2002 midsized vans.

The part was stiffer, lighter, less brittle in cold temperatures, and more recyclable. Molded of a thermoplastic olefin clay, nanocomposite offers weight savings and improved surface quality. GM uses approximately 540,000 lb of nanocomposite material per year. GM research and development worked with Basell, the supplier of polyolefins, and Southern Clay Products, the supplier of nanoclay. The technology for dispersing the clay came from Basell and Southern Clay Products supplied the Impala project with Cloisite nanoclay, i.e., organically modified montmorillonite. Organically modified nanometer scale layered magnesium aluminum silicate platelets measuring 1 nm thick and 70 to 150 nm across were used as additives. Foster Corporation manufactures selectively enhanced polymer (SEP), a line of nanocomposite nylon. Demand for nanocomposite nylon has grown in medical device applications, especially in the catheter tube area. Foster president Larry Aquorlo introduced the product in 2001. Made from Nylon 12 and nanoclay particles, one of the dimensions is less than 1 nm and the other is 1500 times longer. It offers 65 percent increase in flexural modulus, 135 percent elongation of the material with increased stiffness, and rigidity without brittleness. Triton systems won a 5-year contract from the navy to develop application for its Nanotuff, an advanced abrasion resistant and chemical resistant coating for transparent substrates. Procured from nanometer-sized particles suspended in an epoxy-containing matrix, the coating is 4 times more durable than other conventional coatings. Surfaces coated with Nanotuf do not craze, crack, or shatter upon impact and are flame retardant. In 2003, the worldwide consumption of polymer nanocomposites reached 24.5 million pounds valued at U.S. $90.8 million.

Miniaturization improves the performance properties of materials. For example, 6-nm copper grains showed 5 times more hardness than conventional copper. Carbon nanotubes and nanotube composite fibers are several times tougher than steel, Kevlar, or spider silk, which are the strongest materials known today. In a similar manner, cadmium selenide (CdSe) can yield any color in the spectrum simply by controlling the size of its constituent grains.

The first technology developed by Nanomaterials Technology, Singapore, which was incorporated in 2000, is for the production of nanosized precipitated calcium carbonate (NPCC). The high gravity controlled precipitation, HGCP platform can produce ultrafine cubic shaped NPCC with properties such as an average particle size 40 nm, narrow particle size distribution, Brunauer, Emmett and Teller, BET surface area >40 m^2/g, and no chemical inhibitor. In addition to NPCC, NanoMaterials Technology (NMT) has also developed nanoelectronic materials. Nanosized powders such as $BaTiO_3$ enable multilayer ceramic capacitor (MLCC) manufacturers to minimize their electrode layer thickness, hence increasing their volume loadings, resulting in greater efficiency and a lower production cost. NMT targets the following areas in the pharmaceutical sector:

1. *Poorly water-soluble drugs*—A reduction in particle size of poorly water-soluble drugs will greatly increase the surface area of the active ingredient, which will result in a substantial increase in the dissolution rate. The reduction in particle size also enables more stable suspensions to be prepared.

2. *Inhalation drugs*—The HGCP technology platform allows control of particle size and morphology, hence enabling us to have complete control of the particle design for different delivery systems.

3. *Drug carriers*—The HGCP technology platform can also be used to control the synthesis of nanodrug carriers.

Altair Nanotechnologies, Reno, Nevada received an order for preparing Li-ion battery electrodes using nanomaterials last year. The order was from Advanced Battery Technologies for 1 metric ton of lithium titanate spinel electrode nanomaterials. The 2200 lb (1000 kg) of anode electrode nanomaterials, manufactured at Altairnano's facilities in Reno, were to be shipped in September 2005 for Advanced Battery's development program for polymer lithium-ion (PLI) batteries for electric vehicles. The developmental PLI batteries were to be tested in standard battery test protocols and in real-world road tests in one electric bus and one electric sedan. Altairnano is a leading supplier and innovator of advanced ceramic nanomaterial technology. Based in Reno, Nevada, Altairnano has assembled a unique team of materials scientists who, coupled in collaborative ventures with industry partners and leading academic centers, have pioneered an array of intellectual property and products. Altairnano's robust proprietary technology platforms produce a variety of crystalline and noncrystalline nanomaterials of unique structure, performance, quality, and cost. The company has scalable manufacturing capabilities to meet emerging nanomaterial demands, with the production capacity of hundreds of tons of nanomaterials.

Altairnano's two divisions, Life Sciences and Performance Materials, are focused on applications where its nanotechnology may enable new high growth markets. The Life Sciences Division is pursuing market applications in pharmaceuticals, drug delivery, dental materials, and other medical markets. The Performance Materials Division is focused on market applications in advanced materials for paints and coatings and air, water treatment, and alternative energy including new Li-ion battery electrode materials. Advanced Battery Technologies, founded in September 2002, develops, manufactures, and distributes rechargeable PLI batteries. Advanced Battery's products include rechargeable PLI batteries for electric automobiles, motorcycles, mine-use lamps, notebook computers, walkie-talkies, and other personal electronic devices. Advanced Battery's batteries combine high-energy chemistry with state-of-the-art polymer technology to overcome many of the shortcomings associated with other types of rechargeable batteries. The company has an office in New York, with manufacturing facilities in China.

1.2 Feynman's Vision—There's Plenty of Room at the Bottom

Richard Feynman, Nobel laureate, gave an after-dinner talk on December 29, 1959 at the annual meeting of the American Physical Society at the California Institute of Technology. He talked about the problem of manipulation and control of things on a small scale. He drew examples from the endeavors of Kamerling Onnes in the field of low temperature physics, Percy Bridgman and design of higher and higher pressure systems in highlighting the benefits of pushing what good engineers and technologists do to the extreme limit that scale would permit. He discussed the progress made by miniaturization. Electric motors the size of a fingernail and the text of the Bible stored on the device the size of a pinhead were emerging from the laboratory. For example, if the area of all the pages of *Encyclopedia Britannica* is diminished 25,000 times, then a reduced area the same as that of the head of a pin with 1/16-in diameter would emerge. The resolution power of the eye is approximately 1/120 in and when demagnified 25,000 times becomes 80 A in diameter. This would be 32 atoms across. The way this can be achieved is to press the metal into a plastic material and form a mold, then peel the plastic, evaporate the silica, and form a thin film. Then shadow it by evaporation of gold at an angle against the silica so that all the little letters will be clear, followed by dissolution of the plastic away from the silica film. Then observe it with an electron microscope.

Feynman discussed methods of writing small. He felt that it was not as difficult as it first appeared to be. The lenses of the electron microscope when reversed demagnify. A source of ions can be focused on a very small spot and sent through the microscope lens in reverse. Writing can be affected with the ions much as writing on a TV cathode ray oscilloscope. The amount of material that is deposited can be adjusted. This process may be slow. Rapid methods may have to be developed. Feynman suggested the use of photo with a screen that has holes in it in the form of letters. A system of lenses can be used to form images using metallic ions that would deposit the metal on the pin. He alluded to a simpler method using light, optical microscope, and by focus of electrons on a small photoelectric screen. The beam may etch away the metal when run long enough. He considered all the books in the world, i.e., 24 million volumes. The pins needed would lie within 3 square yards. Thus, libraries filled with books could be stored in something the size of a library card that can be held in one's palm. He also considered further shrinkage of size needed for storage of all the information available in all the books in the world. This can be achieved by using codes such as dots and dashes to represent all the letters. Allowing 100 atoms for each bit of information, he calculated that in a cube of material 1/200 in width is sufficient for such storage. This is the barest piece of dust that can be resolved by the human eye. Further, he pointed out that storage of enormous amounts of information in exceedingly small space is nothing new to biologists. DNA molecules, where 50 atoms are used for one bit of information about the cell, are where all the information about all the functions of any given organism is stored. The molecules are 2.5 nm in diameter.

He called for design and development of better electron microscopes. The electron microscope in those days could resolve only approximately 10 A. As the wavelength of the electron is only 1/20 of an Angstrom, the electron microscope can be made with 100 times more resolution power. Atoms can be viewed individually using such microscopes. Sequences of bases in DNA were studied. Today it has emerged as the field of bioinformatics. The protein synthesis and DNA transcription, translation, and replication are better understood at this time and age. Feynman talked about the numerical aperture and f value of lenses. Although there are theorems that prove it is impossible to seek an increase in numerical aperture above a certain value, he called for electron microscopes to be made more powerful.

He marveled at the smallness of the biological system. Many of the cells are tiny and active. They produce different substances and they walk around. In computers, information is written, erased, and overwritten in a small space. He posed the question why computers that fill several rooms could not be made smaller. For instance, wires can be 100 atoms in diameter and the circuits can be a few thousand Angstroms in width. Computers can be made to be more complicated by several orders of magnitude. They could make judgments when they have millions of times as many elements. They can recognize faces. Feynman wanted elements of the computer to be made submicroscopic. If a computer is made with all the features desired, it may be of the size of the Pentagon and may require all the Germanium material available in the world. Problems of heat generation and power consumption remain. He called for miniaturization of computers, citing the speed of light. Increase in the speed of computers could be achieved upon miniaturization, as information has to travel from one location to another within the computer.

Feynman alluded to a process of evaporation and formation of layered materials much like the atomic layer deposition methods used currently. He called for drilling holes, cutting things, soldering things, stamping things out, and molding different shapes at an infinitesimal level. Miniaturization of automobiles may need larger tolerances. With decrease in scale some problems of lubrication becomes apparent. The effective viscosity of oil would be higher. A wild idea is to swallow the surgeon during surgery. A mechanical surgeon inside the blood vessel can go into the heart and look around. It detects the faulty valve and slices it out using a knife. He discussed a master-slave system that operated electrically. Slaves are made by large-scale machinists so that they are one-fourth the scale of the "hands." The little servo motors with little hands play with little nuts and bolts, drill holes, and are 4 times smaller. For example, when a quarter size lathe is manufactured, another set of hands are made one-fourth size. When finished they are wired to the large-scale system through transformers to one-sixteenth size servo motors. A pantograph can be used to achieve a factor of four reduction. A pantograph can make a smaller pantograph that can make a smaller pantograph and so on. Flats can be made by rubbing unflat surfaces in triplicate together. The flats become flatter than the thing with which we started. Precision is improved on a small scale. A baby lathe 4000 times smaller than usual is achieved. Feynman pondered as to how many washers can be made by this lathe for the computer that is going to be made by drilling holes using the baby lathe.

With 2 percent of material used to make a big lathe, a billion little lathes each 1/4000 scale of a regular lathe can be made. These lathes can be used to drill holes, stamp parts, and so on. Many things do not scale down in proportion. Materials may adhere to each other by van der Waals forces of attraction.

Can atoms be rearranged at will? When elements are beneficiated from ores, the impurities are removed. Can materials be devised with atoms arranged the way we want them? What are the problems associated with building electric circuits in small scale? As the wavelength goes down with scale, the natural frequency goes up. The problem of resistance remains. The problem of resistance can be overcome by superconductivity if the frequency is not too high.

When a circuit of seven atoms is considered, new opportunities for design open up. Quantum mechanics is more applicable to describe the world of seven atoms compared with the macroworld. The devices can be mass-produced. Larger machines cannot be made as exact copies of each other. New kinds of forces and new kinds of possibilities open up at the atomic level. Problems of manufacture and reproduction of materials will be different. Physicists can synthesize any chemical substance that the chemist writes down. Feynman called for a competition where one laboratory makes a tiny motor and sends it to another laboratory where they manufacture a thing that fits inside the shaft of the first motor. He suggested a competition between high schools, sending pins to each other and working on it.

He offered $1000 to the first person who could take the information on the page of a book and put it on an area 1/25,000 smaller in linear scale in such a fashion that it can be read using an electron microscope. Another $1000 was offered to anyone who could make an operating electric rotating motor, 1/64 in cube, which could be controlled from the outside. The 1/64 in cube dimensions were not counting the lead-in wires. He did not expect the prize to wait very long for claimants.

1.3   Drexler-Smalley Debate on Molecular Assemblers

R.E. Smalley and K.E. Drexler debated on whether "molecular assemblers," which are devices capable of positioning atoms and molecules for precisely defined reactions in any environment, are possible. In his book *Engines of Creation: The Coming Era of Nanotechnology*, Drexler envisioned a world ubiquitous with molecular assemblers. These would provide immortality and lead to the colonization of the solar system. He received a Ph.D. from Massachusetts Institute of Technology (MIT) in 1991. He is also the CEO of Foresight Institute, Palo Alto, California. Smalley, a recipient of the Nobel Prize in chemistry in 1996 for his work on fullerenes, outlined his objections based on science to the molecular assembler idea and called it the "fat fingers problem" or the "sticky fingers problem." He was also worried about funding for nanotechnology due to the portrayed darker side of it. In *Chemical and Engineering News*, "Point–Counterpoint" column, an open letter from Drexler to Smalley was posted challenging Smalley to clarify the "fat fingers problem," with a response from Smalley and three letters with Drexler countering and Smalley concluding the exchange.

Drexler sought clarification from Smalley on the fat fingers problem. He felt that like enzymes and ribosome the proposed molecular assemblers neither have nor need the "Smalley Fingers." Drexler alluded to the long-term goal of molecular manufacturing and its consequences, which can pose opportunities and dangers to long-term security of the United States and the world. Theoretical studies and implementation capabilities are akin to the pre-Sputnik studies of spaceflight or the pre-Manhattan project calculations regarding nuclear chain reactions and are of more than academic interest. He referred to his 20-year history of technical publications in the area of chemical synthesis of complex structures by mechanically positioning reactive molecules and not by manipulating individual atoms. The proposal was successfully defended in the doctoral thesis on *Nanosystems: Molecular Machinery, Manufacturing, and Computation* and is based on well-established physical principles.

Smalley responded with an apology should he have offended Drexler in his article in *Scientific American* in 2001. He called for agreement that the precision picking and placing of individual atoms using "Smalley Fingers" is impossible. A "Smalley Finger" type of molecular assembler tool will never work. Smalley pointed out the infeasibility of tiny fingers placing one atom at a time. This is also applicable to placing larger, more complex building blocks. As each incoming reactive molecule building block has multiple atoms to control during the reaction, more fingers are needed to ensure they do not go astray. Computer-controlled fingers will be too fat and too sticky for providing the control needed. Fingers cannot perform the chemistry necessary. He called attention to the mention of enzymes and ribosomes needed in the reaction medium. He quarreled with the vision of a self-replicating nanobot. Is there a living cell inside the nanobot that cranks these out? Water is needed inside the nanobot with the necessary nutrients for life. How does the nanobot pick the right enzyme and join in the right fashion? How does the nanobot perform error detection and error correction? He worried about the scope of the chemistry that the nanobot could perform. Enzymes and ribosomes need water to be effective. He mentioned that although biology is wondrous, a crystal of silicon, steel, copper, aluminum, titanium, and other key materials of technology could not be produced by biology. Therefore, without these materials how could a nanobot manufacture a laser, ultrafast memory, and other salient components of modern society?

Drexler applauded Smalley's goal of debunking nonsense in nanotechnology. He sketched the fundamental concepts of molecular manufacturing. He referred to Feynman's after-dinner visionary talk in 1959, discussed in Sec. 1.2, and nanomachines building atomically precise products. Feynman's nanomachines were largely mechanical and not biological. In order to understand how a nanofactory system could work, he considered a conventional factory system. Some of Smalley's questions reach beyond chemistry to systems engineering and to problems of control, transport, error rates, and component failure and to answers from computers, conveyors, noise margins, and failure-tolerant redundancy. Nanofactories contain no enzymes, no living cells, and no replicating nanobots, but they do use computers for precise control, conveyors for parts transport, and positioning devices of assorted sizes to assemble small parts into large parts when building macroscopic products. The smallest devices position molecular parts to assemble structures through mechanosynthesis or machine-based chemistry. Conveyors and positioners bring reactants together unlike solvents and thermal motion. Positional control enables a strong catalytic effect by aligning reactants for repeated collisions in optimal geometries at vibrational frequencies greater than terahertz. Positional control can lead to voiding unwanted side reactions. From transition state theory, for suitably chosen reactants, positional control will enable synthetic steps at megahertz frequencies with reliability approaching that of digital switching operations in a computer. When molecules come together and react, their atoms, being sticky, stay bonded to neighbors and thus do not need sticky fingers to hold them. Direct positional control of reactants is revolutionary and achievable. Mechanosynthetic reactions and its field have parallels in the field of computational chemistry. The flourishing of nanotechnology in 2003 suggests a bottom-up strategy using self-assembly. It used to be scaling down microscopic machines in 1959. Progress toward molecular manufacturing is achieved by research in computational chemistry, organic synthesis, protein engineering, supramolecular chemistry, and scanning probe manipulation of atoms and molecules. Scaling down moving parts by a factor of one million results in multiplication of their frequency of operation by the same factor. Progress in the United States on molecular manufacturing has been impeded because of the illusion that it is infeasible. He called for augmentation of nanoscale research with a systems engineering effort and achievement of the grand vision articulated by Richard Feynman.

Smalley concluded by observing that Drexler left the talk about real chemistry and went to the mechanical world. He felt that precise chemistry could not be made to happen as desired between two molecular objects with simple mechanical motion along a few degrees of freedom in the assembler fixed frame of reference. It was agreed that a reaction would be obtained when a robot arm pushes the molecules together but it may not be the reaction desired. More control is needed than mentioned about molecular assemblers. A molecular chaperone is needed that serves as catalyst. Some agent such as an enzyme is needed. A liquid medium such as water is needed to complete the desired chemical reactions. Smalley recalled a talk on nanotechnology he gave to 700 middle and high school students. The students were asked to write an essay on "Why I am a Nanogeek." Smalley read the top 30 essays and he picked his favorite five. Half assumed that self-replicating nanobots were possible and some were worried about if filled the world. However, they have been misinformed.

1.4   Chronology of Events during the Emergence of Nanotechnology

The chronology of events that are significant in the emergence of the field of nanotechnology is shown in Table 1.1. Feynman's after-dinner talk and the Drexler–Smalley debate on feasibility of molecular assemblers and self-replicating nanobots in 2003 were discussed in the previous sections.

1959	Feynman gives after-dinner talk describing molecular machines building with atomic precision http://www.zyvex.com/nanotech/feynman.html
1974	Taniguchi uses the term Nanotechnology in journal article on ion-sputter machining http://www.nanoword.net/library/nwn/1.htm
1977	Drexler originates molecular nanotechnology concepts at MIT
1981	STM invented First technical paper on molecular engineering to build with atomic precision http://www.imm.org/PNAS.html
1985	Buckyball discovered
1986	*Engines of Creation* by Eric K. Drexler, first book published AFM invented First organization formed
1987	First protein engineered First university symposium
1988	First university course http://www.foresight.org/Updates/Update18/Update18.1.html#FirstCourse
1989	IBM logo spelled in individual atoms First national conference
1990	First nanotechnology journal Japan's STA begins funding nanotech projects
1991	"Bottom-up" path selected Japan's MITI announces atom factory IBM endorsement Carbon nanotube (CNT) discovered
1992	*Nanosystem: Molecular Machinery, Manufacturing & Computation* by Eric K. Drexler, first textbook published First congressional testimony
1993	First Feynman Prize in nanotechnology awarded First coverage of nanotech from White House *Engines of Creation* book given to Rice administration, stimulating first university nanotech center
1994	Nanosystems textbook used in first university course U.S. science advisor advocates nanotechnology

TABLE 1.1   Chronology of Events during Emergence of Nanotechnology

1995	First think tank report
	First industry analysis of military applications
1996	$250,000 Feynman grand prize announced
	First European conference
	NASA begins work in computational nanotech
	First nanobio conference
1997	First company founded: Zyvex Instruments, Richarson, TX
	First design of nanorobotic system
1998	First National Science Foundation, NSF Forum, held in conjunction with Foresight Conference
	First DNA-based nanomechanical device
1999	First nanomedicine book published
	First safety guidelines
	Congressional hearings on proposed national nanotechnology initiative
2000	President Clinton announces U.S. National Nanotechnology Initiative (NNI)
	First state research initiative: $100 million in California
2001	First report on nanotech industry
	United States announces first center for military applications
2002	First nanotech industry conference
	Regional nanotech efforts multiply
2003	Congressional hearings on societal implications
	Call for balancing NNI research portfolio
	Drexler–Smalley debate is published in *C&E News*
2004	First policy conference on advanced nanotech
	First center for nanomechanical systems
2005	At nanoethics meeting, NSF announces nanomachine/nanosystem, project count has reached 300

TABLE 1.1   Chronology of Events during Emergence of Nanotechnology (*Continued*)

1.5   Applications

Nanotechnology is a corollary from the computer revolution. The gate width of microprocessors is made smaller and smaller as the speed of the microprocessor increases according to Moore's law. According to the laws of physics, the minimum gate width achievable is approximately 10 nm. At the speed of light should the CPU signal traverse the gate, the reciprocal of time taken, the frequency of the process would be $3.0 \ 10^8 \, \text{m/s} / 10^{-8} = 3 \times 10^{16}$ Hz. The miniaturization of the computer and onslaught of silicon chips and packing of several transistors on the chip is one important application of nanotechnology.

Antibacterial wound dressings use nanoscale silver. A nanoscale dry powder can neutralize gas and liquid toxins in chemical spills and in other places. Batteries manufactured with nanoscale materials can deliver more power with increased speed and with less heat generation. Cosmetics and food producers are miniaturizing their ingredients to the nanoscale in order to improve their effectiveness. Sunburns can be prevented using sunscreens that contain nanoscale titanium dioxide or zinc oxide. Scratch and glare-resistant coatings have been developed using nanotechnology, enabling glasses to be UV proof and transparent. Drug delivery techniques are improved using nanotechnology. Dendrimers are nanostructures that can be precisely designed and manufactured for applications including treatment of cancer and other diseases. Diagnosis of disease states can be facilitated by nanotechnology. Nanofilms can be used on eyeglasses, computer displays, and cameras to protect or treat surfaces.

CNTs can be used to increase the photovoltaic efficiency of solar cells. This can reduce our dependence on oil imports and result in the use of renewable energy resources. CNTs can be used in baseball bats, tennis racquets, and automobile parts. The use of CNTs in novel materials results in increased strength at reduced weight. CNTs posses very high thermal conductivity and can be used well in thermal management. Interesting electronic properties of CNTs enable their use for flat panel displays in television sets, thermal batteries, and other electronics. Nanotube synthesis from materials other than carbon is also contemplated. Transistors that are used for electronic switching decrease in size per Moore's law, resulting in greater power for the computer. Current-day silicon chips possess features as small as 45 nm. Nanoscale materials absorb solar light and convert them to electrical energy in a more efficient fashion. Thin-film solar cells paired with a new kind of rechargeable battery also are currently being researched. CNT-based membranes can be used in seawater desalination. Nanoscale sensors can be used to identify contaminants in water systems.

Nanoparticles are being used in a number of industries. Nanoscale materials are used in electronic, magnetic, optoelectronic, biomedical, pharmaceutical, cosmetic, energy, catalytic, and materials applications. Areas producing the greatest revenue for nanoparticles reportedly are chemical-mechanical polishing, magnetic recording tapes, sunscreens, automotive catalyst supports, biolabeling, electroconductive coatings, and optical fibers.

At this time, most computer hard drives contain giant magneto resistance (GMR) heads that through nanothin layers of magnetic materials allow for an order of magnitude increase in storage capacity. Other electronic applications include nonvolatile magnetic memory, automotive sensors, landmine detectors, and solid-state compasses. Nanomaterials, which can be purchased in dry powder form or in liquid dispersions, often are combined with other materials to improve product functionality. Additional products available today that benefit from the unique properties of nanoscale materials include step assists on vans, bumpers on cars, paints and coatings to protect against corrosion, scratches and radiation, protective and glare reducing coatings for eyeglasses and cars, metal-cutting tools, sunscreens and cosmetics, longer-lasting tennis balls, light-weight stronger tennis racquets, stain-free clothing and mattresses, dental bonding agents, burn and wound dressings, and automobile catalytic converters.

With basic research underway for 20 plus years, nanotechnologies are gaining in commercial introductions. In the short term, nanoparticles will be introduced into many existing materials, making them stronger or changing their conductive properties. Significantly, stronger polymers will make plastics more widely used to reinforce materials and replace metals, even in the semiconductor area.

One of the most innovative products is one that enhances biological imaging for medical diagnostics and drug discovery. Quantum dots are semiconducting nanocrystals that, when illuminated with UV light, emit a vast spectrum of bright colors that can be used to identify and locate cells and other biological activities. These crystals offer optical detection up to 1000 times brighter than conventional dyes used in many biological tests, such as MRIs, and render significantly more information. A recent display technology for laptops, cell phones, and digital cameras is made of nanostructured polymer films. Known as organic light emitting diodes (OLEDs), several large companies will begin producing them. Among OLED screen advantages are brighter images, lighter weight, less power consumption, and wider viewing angles. Nanoparticles also are being used increasingly in catalysis where the large surface area per unit volume of nanosized catalysts enhances reactions. Greater reactivity for these smaller agents reduces the quantity of catalytic materials necessary to produce the desired results. The oil industry relies on nanoscale catalysts for refining petroleum, while the automobile industry is saving large sums of money by using nanosized—in place of larger—platinum particles in its catalytic converters. Because of their size, filters made of nanoparticles also have been found to be excellent for liquid filtration. Several products are now available for large-scale water purification that can remove the tiniest bacteria and viruses from water systems in addition to chemicals and particulate matter.

Another application of nanotechnology in useful applications is in the field of wear resistant coatings. In the mid-1990s, nanoceramic coatings exhibiting much higher toughness than conventional coatings were developed. Beginning in 1996, the Department of Defense (DoD) supported partnerships among the Navy, academia, and industry to develop processes suitable for use in manufacturing and to evaluate the coatings for use in marine environments. In 2000, the first nanostructural coating was qualified for use on gears of air-conditioning units for U.S. Navy ships. In 2001, the technology was selected to receive an R&D 100 award. DoD estimates that use of the coatings on air valves will result in a $20 million reduction in maintenance costs over 10 years. The development of wear-resistant coatings by the DoD is clearly allied with its mission, yet will lead to commercial applications that can extend the lifetime of moving parts in everything from personal cars to heavy industrial machinery. The pharmaceutical and chemical industries are being affected greatly by nanotechnology. New commercial applications of nanotechnology expected in these and other industries include:

- advanced automatic drug delivery system, implantable devices
- medical diagnostic tools, cancer tagging mechanisms, lab-on-chip
- cooling chips or wafers to replace compressors in cars, appliances
- sensors for airborne chemicals
- photovoltaics, solar cells, fuel cells, portable power
- new high performance materials

It is difficult to predict what products will move from the laboratory to the marketplace over longer periods. However, it is believed that nanotechnology will facilitate the production of ever-smaller computers, store vastly greater amounts of information, and process data much more quickly than those available today. Computing elements are expected to be so inexpensive that they can be in fabrics for smoke detection for instance and other materials.

The earliest application of nanoscale materials occurred in systems where nanoscale powders could be used in their free form, without consolidation or blending. For example, cosmetics manufacturers now commonly use nanoscale titanium dioxide and zinc oxide powders for facial base creams and sunscreen lotions. Nanoscale iron oxide powder is now being used as a base material for rouge and lipstick. Paints with reflective properties are also being manufactured using nanoscale titanium dioxide particles. Nanostructured wear-resistant coatings for cutting tools and wear-resistant components have been in use for several years. Nanostructured cemented carbide coatings are used on some navy ships for their increased durability.

Recently, uses of nanoscale materials that are more sophisticated have been realized. Nanostructured materials are in wide use in information technology, integrated into complex products such as hard disk drives that store information and the silicon integrated circuit chips that process information in every Internet server and personal computer. The manufacture of silicon transistors already requires the controlled deposition of layered structures just a few atoms thick (approximately 1 nanometer). Lateral dimensions are as small as 180 nm for the critical gate length, and semiconductor industry roadmaps call for them to get even smaller. With shorter gate lengths come smaller, faster, more power-efficient transistors and corresponding improvements in the cost and performance of every digital appliance. Similar processes are required for the manufacture of information storage devices. The GMR read heads in computer industry standard hard disk drives are composed of carefully designed layered structures, where each layer is just a few atoms thick. The magnetic thin film on the spinning disk is also a nanostructured material. IBM recently announced the introduction of an atomically thin layer of ruthenium (humorously referred to as "pixie dust") to substantially increase the information storage density of its products. Greater storage density translates directly to the less expensive storage of information. Incorporating nanostructured materials and nanoscale components into complex systems, both magnetic data storage and silicon microelectronics provide a glimpse of the future of nanoscale science and technology.

In biomedical areas, structures called liposomes have been synthesized for improved delivery of therapeutic agents. Liposomes are lipid spheres approximately 100 nm in diameter. They have been used to encapsulate anticancer drugs for the treatment of AIDS-related Kaposi's sarcoma. Several companies are using magnetic nanoparticles in the analyses of blood, urine, and other body fluids to speed up separation and improve selectivity. Other companies have developed derivatized fluorescent nanospheres and nanoparticles that form the basis for new detection technologies. These reagent nanoparticles are used in new devices and systems for infectious and genetic disease analysis and for drug discovery. Many uses of nanoscale particles have appeared in specialty markets, such as defense applications, and in markets for scientific and technical equipment. Producers of optical materials and electronics substrates such as silicon and gallium arsenide have embraced the use of nano-size particles for chemo-mechanical polish.

1.6   Nanotechnology Challenges

1.6.1   Fundamental Physical Limits of Miniaturization

The history of information technology has been largely a history of miniaturization based on a succession of switching devices, each smaller, faster, and cheaper to manufacture than its predecessor. The first general-purpose computers used vacuum tubes, but the tubes were replaced by the newly invented transistor in the early 1950s, and the discrete transistor soon gave way to the integrated circuit approach. Engineers and scientists believe that the silicon transistor will run up against fundamental physical limits to further miniaturization in perhaps as little as 10 to 15 years, when the channel length, a key transistor dimension, reaches something like 10 to 20 nm. Microelectronics will have become nanoelectronics, and information systems will be far more capable, less expensive, and more pervasive than they are today. Nevertheless, it is disquieting to think that today's rapid progress in information technology may soon end. Fortunately, the fundamental physical limits of the silicon transistor are not the fundamental limits of information technology. The smallest possible silicon transistor probably will still contain several million atoms, far more than the molecular-scale switches that are now being investigated in laboratories around the world. However, building one or a few molecular-scale devices in a laboratory does not constitute a revolution in information technology. To replace the silicon transistor, these new devices must be integrated into complex information processing systems with billions and eventually trillions of parts, all at low cost. Fortunately, molecular-scale components lend themselves to manufacturing processes based on chemical synthesis and self-assembly. By taking increasing advantage of these key tools of nanotechnology, it may be possible to put a cap on the amount of lithographic information required to specify a complex system, and thus a cap on the exponentially rising cost of semiconductor manufacturing tools. Thus, nanotechnology is probably the future of information processing, whether that processing is based on a nanoscale silicon transistor manufactured to tolerances partially determined by processes of chemical self-assembly or on one or more of the new molecular devices now emerging from these laboratories. Nanosize particles of silicon carbide, diamond, and boron carbide are used as lapping compounds to reduce the waviness of finished surfaces from corner to corner and produce surface finishes of 1 to 2 nm smoothness. The ability to produce such high-quality components is significant for scientific applications and could become even more important as electric devices shrink and optical communications systems become a larger part of the nation's communications infrastructure.

1.6.2 Thermodynamic Stability of Nanostructures[1]

The solvation thermodynamics define the stability of a system. It can be stable, metastable, or unstable. Upper critical solution temperature, UCST and lower critical solution temperature, LCST define the critical consolute temperatures of two-phase systems. When a supersaturated system is disturbed, the particles begin to nucleate, grow, and form stable structures. The process can be arrested sufficiently early to form a nanosphere. However, the free energy of formation of the structure and the surface energy of the solid can be equated with each other at equilibrium. For stability, the free energy has to be negative or equal to zero. From these criteria, a minimum stable size of the solid particle formed can be calculated. This depends on the solid–liquid surface tension values and other parameters of the system. In one system, for example, the engineering thermoplastic Acrylonitrile-Butadiene-Styrene (ABS) the smallest stable butadiene particle size can be calculated as 200 nm from Gibbs free energy. When making smaller particles is attempted, the rubber phase particles agglomerated with each other to a size larger than 200 nm. It was reported by a number of investigators that making the rubber particles smaller and smaller was difficult. Maybe if they were made into tubes, they could be made into nanotubes. The free energy and surface energy analysis will still hold. The shape is different in the derivation.

The four thermodynamically stable forms of carbon are diamond, graphite (C_{60}), buckminsterfullerene, and CNT. It would be a challenge to extend the experience gained in CNT to nanotubes made of material other than carbon. It would also be interesting to form stable spherical structures in the nanoscale dimensions without agglomeration. At what scale would the quantum analysis for atoms be applicable when compared with the Newtonian mechanics used to describe macrosystems? Nanostructures of all different shapes and Bravais lattices in several materials need to be established. Nanostructures that are known today and successfully used in the industry are in the form of tubular morphology, gate patterns of oxidation, and packing transistors that leave some features at the nanoscale dimensions. Why is tubular morphology favored over spherical morphology during the formation of CNTs? The layered materials of the nanoscale dimension made using atomic layer deposition techniques break no known laws of thermodynamics. However, one issue is the layer rearrangement due to Marangoni instability.

1.6.3   Characterization of Nanostructures

Characterization of nanostructures, per se, is at more resolution than optical microscopes. X-rays and neutron-based devices are needed. The scanning electron microscope, SEM, scanning probe microscope, SPM, and atomic force microscope, AFM are progress in this area. Feasibility needs to be established for molecular manufacturing and self-replicating nanobots as Smalley points. According to the Raleigh criterion, the minimum resolution size achievable is half the wavelength of light, 200 nm. Small angle x-ray scattering, SAXS and wide angle x-ray scattering, WAXS are needed to study nanomaterials in the laboratory.

1.7   Summary

Synthesis, characterization, and applications of matter with atleast one dimensions less than 100 nm is the field of nanotechnology. Raleigh criterion sets the resolution size at 200 nm. Vision precedes invention. Commercial products have been introduced such as nano-porous catalysts, clay-filled nylon composites, intercalated polyole-fin nanocomposites, nanocomposite nylon, novel coatings, CNTs carbon nanotubes, nanoelectrodes in thermal batteries, and so on.

Richard Feynman's after-dinner talk in 1959 provided the vision for storage of an encyclopedia within the size of a pinhead, the method of writing small using ions, and the focus of electrons on a small photoelectric screen. He called for the design and development of better electron microscopes with the capability to view atoms, better f value lenses, making smaller computers that used to fill several rooms, and elements of the computer to be made submicroscopic. He alluded to a process of evaporation and formation of layered materials much like atomic layer deposition, ALD methods used currently. He called for drilling holes, cutting things, soldering things, stamping things out, and molding different shapes at an infinitesimal level. A pantograph can make a smaller pantograph that can then make a smaller pantograph and so on. He mused whether atoms can be rearranged at will. He offered $1000 to the first person who could take the information on the page of a book and put it in the area 1/25,000 smaller in linear scale in such a fashion that it could be read using an electron microscope.

Drexler and Smalley debated whether molecular assemblers, devices that are capable of positioning atoms and molecules for precisely defined reactions in any environment, are possible. Smalley objected to the immortality implied by Drexler's vision of a molecular assembler in the form of the "fat fingers problem." *Chemical and Engineering News* carried the post of Drexler to Smalley with response from Smalley by three letters with Drexler countering and Smalley concluding the exchange on the sticky fingers problem. Drexler alluded to the long-term goal of molecular manufacturing. Smalley sought agreement that precision picking and placing of individual atoms using Smalley fingers is impossible. Computer-controlled fingers would be too fat and too sticky for providing the control needed. Enzymes and ribosomes were necessary to complete the reactions. Smalley quarreled with the vision of the self-replicating nanobot. Water is needed to provide the nutrients. Drexler applauded Smalley's goal of debunking nonsense in nanotechnology. Direct positional control of reactants is revolutionary and achievable.

The chronology of events that mark the rise of nanotechnology as a discipline is shown in Table 1.1, from Feynman's talk in 1959 to the nanoethics meeting in 2005. A wide range of applications is expected in nanotechnology ranging from solar cells with increased photovoltaic efficiency, to sunscreens to GMR, to giga magnetic hardrives. The challenges in nanotechnology that need to be overcome are fundamental physical limits to miniaturization, thermodynamic stability of nanostructures and existence of a minimum size below which spheres tend to agglomerate, layer arrangement, and why tubular morphology is preferred to spherical morphology in the nanoscale range. Some characterization tools needed in nanotechnology are SEM, AFM, SAXS, and WAXS.

Review Questions

1. Find the side of the cube that contains 1 attomole of silver.

2. How many grams of silica are present in a sphere with a diameter of 5 nm?

3. How many molecules of copper are present in a cube of side 50 nm?

4. Who invented the field of low temperature?

5. What was Percy Bridgman's claim to fame?

6. Explain the Raleigh criterion for minimum resolution size.

7. What did Feynman mean by miniaturization of the computer?

8. Does a circuit of seven atoms satisfy the laws of quantum mechanics?

9. How would you define nanotechnology?

10. What is a nanostructure?

11. When is something called a nanodevice?

12. What is the ion microscope?

13. Why is a scanning probe microscope needed?

14. What is the diameter of an electron?

15. Can atoms be rearranged at will?

16. How many lathes can be made with 2 percent of the material from a regular size lathe?

17. What is a pantograph?

18. What did Feynman illustrate by taking lubrication as an example?

19. What is miniaturization by evaporation?

20. What is the f value in the lens of an electron microscope?

21. What is the resolving power of the eye?

22. What is the fat fingers or sticky fingers objection of Smalley?

23. What are nanobots and how is it impossible to control the chemistry with them?

24. Discuss the self-replicating nanoassembler.

25. What are the implications of a self-replicating nanoassembler?

26. Why did Drexler discuss enzymes and ribosome in his debate with Smalley?

27. Why cannot molecular assemblers perform complex chemistry?

28. Discuss the use of CNTs in solar cells.

29. Discuss the use of nanostructures in disease diagnosis.

30. What is the role of nanostructures in drug discovery?

31. What is "pixie dust"?

32. What is a quantum dot?

33. What is the fastest computer possible?

34. What is the smallest gate size achievable in silicon chips?

35. What is the difference between mechanosynthetic reactions and traditional chemistry?

36. Can the use of conveyors and positioners of atoms lead to the void of undesired side reactions?

37. What is the role of water in reactions?

38. In his visionary talk what did Feynman have to say about the demagnification concept??

39. Writing using a light beam or ion beam in small letters was suggested by Feynman to whom?

40. How is nanotechnology used in sunscreens?

41. What is the diameter of a DNA molecule and that of a CNT?

42. What is the length of bacterium and the width of a house?

43. What does the movie *Fantastic Voyage* have to do with nanotechnology?

44. Discuss some features of the first commercial nanocomposite products put forward in the market by Toyota and GM.

45. How are thermal batteries improved by the development of nanotechnology?

46. How has nanotechnology been used in paints?

47. When was the first book on nanotechnology published?

48. When was the first journal on nanotechnology introduced?

49. Which U.S. President first allocated research funds for nanotechnology?

50. What is NNI?

51. Are molecular assemblers feasible?

52. Are self-replicating nanobots feasible?

53. What are MLCCs and how are they used in the enhancement of thermal battery performance?

54. What are PLI batteries and how is nanotechnology research improving their performance?

55. Why did Feynman call for improvement in SEM?

56. Can biological systems produce crystalline silicon, copper, or steel?

57. What are dendrimers and where are they used?

58. How is nanotechnology used in GMRs?

59. What is Kapsosi's sarcoma?

60. What is metastability?

61. Can nanotubes be synthesized using material other than carbon?

62. Can particles with spherical morphology be made with less than 5 nm in diameter?

63. What is the difference between metastability and instability?

64. Name two challenges of nanotechnology in the future.

65. What are the issues involved in characterization of nanostructures?

Reference

1. K. R. Sharma, Thermodynamic Stability of Nanocomposites, Interpack 99, Mauii, HI, 1999.

CHAPTER 2

Fullerenes

Learning Objectives

- Discovery of fullerene structure as the third allotrope of carbon
- 12 × 5 feasibility rule from Euler, stability from rotational spectra and quantum angular momentum calculations
- Patented combustion synthesis method from flame of fuel
- Growth of needle crystals of fullerene by counter diffusion
- Method of organic synthesis of C_{60} "soccer ball" structure
- Solar process for manufacture of fullerenes
- Electric arc process for making fullerenes
- Applications

2.1 Discovery

The third allotropic form of carbon, C_{60}: *Buckminsterfullerene*, was discovered serendipitously in 1985 in an experiment designed to unravel the carbon chemistry in red giant stars (Fig. 2.1). The Nobel Prize in chemistry was awarded to Sir Harold W. Kroto, Richard Smalley, and Robert Curl in 1996 for their work in fullerenes. The birth of fullerene science fell in one place over 10 days in autumn of 1985. A team of scientists discovered the self-assembly of C_{60} from a hot nucleating plasma of carbon.[1] Kroto discovered long carbon chain molecules in the interstellar medium.

Smalley was developing a range of advanced chemical physics techniques, culminating in a clever, pulsed, supersonic nozzle-laser vaporization apparatus that enabled small clusters of refractory materials to be created and probed seminally. It was found that C_{60} molecules existed and they self-assembled. The "soccer ball" structure of carbon was obtained using solvent extraction techniques. Iijima[2] discovered CNTs using an arc-discharge evaporation method

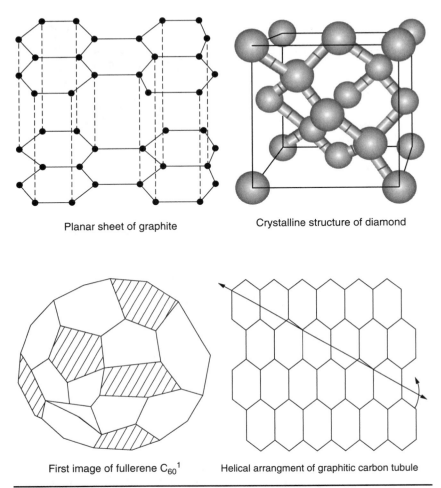

Planar sheet of graphite   Crystalline structure of diamond

First image of fullerene C_{60}[1]   Helical arrangment of graphitic carbon tubule

FIGURE 2.1   Four stable allotropes of carbon—planar sheet of graphite, crystalline diamond, Buckminsterfullerene (C_{60}), and carbon nanotube (CNT).

similar to the method used for fullerene synthesis. The molecule C_{60} is regarded as a beautiful molecule. It possesses incredible symmetry. The molecule resembles the geodesic domes designed by Buckminster Fuller. This molecule has fascinated scientists and humanity of all ages. Kroto showed from quantum angular momentum calculations and rotational molecular spectra that bucky balls are stable structures. Kroto set up a microwave spectroscopy research program that aimed at creation of multiple carbon bonds and various second and third row atoms such as sulfur, phosphorous, and silicon. Long linear chains of carbon atoms were perfect test-beds for quantum mechanical study of the dynamics of bending and rotation in simple linear systems. Their Hewlett-Packard microwave spectrometer enabled the measurement of rotational spectra of molecules.

Ammonia was discovered in Orion. The vast dark clouds between stars consist of methanol, CO_2, formaldehyde, ethanol, HCN, formic acid, formamide, etc. They were detected using radio telescopes and microwave rotational emission spectra. Molecules play a critical role in the collapse of clouds to form stars. Kroto provided a stellar solution to the chain problem of red giant carbon stars. Kroto, Smalley, and Curl detected linear molecules with 5 to 9 atoms of carbon and it became clear that chains originated in the red giant stars.

They detected molecules with 60 carbon atoms and 70 carbon atoms. C_{60} appeared to be unreactive. The flat hexagonal sheets had either formed or ablated from the surface of the graphite disc and closed into a cage, eliminating the reactive edge. The closed hexagonal cage hypothesis reminded Kroto of the visit of his family to Expo67 in Montreal where Buckminster Fuller's dome had dominated the horizon. It had 60 vertices.

Experiments were initiated at Rice University in Houston, Texas. A team of dedicated students constructed a cluster beam machine and the cluster behavior was probed in detail. As all carbon atoms in C_{60} are equivalent, the $^{13}C_{NMR}$ spectrum can be expected to consist of a single elegant line. Kroto presented the conjectured structure in Riccione. Euler's law states that no sheet of hexagons could close. However, if 12 pentagons were introduced into a hexagonal sheet of any size it would close. Further unsaturated molecules with adjacent pentagonal rings were extremely unstable. $5 \times 12 = 60$. Thus, C_{60} had to be the smallest cage able to close without abutting pentagons. The stability of C_{60} requires Euler's 12-pentagon closure principle and the chemical stability conferred by pentagon nonadjacency. Kroto calls it the "2 + 2" of fullerene stability. Magic numbers of smaller cages such as 60, 50, and 28 can be deduced from isolation principle. No cage can be constructed with 22 atoms or fewer. Semistable fullerenes down to C_{20} were predictable. Another investigator attempted to build C_{240}, C_{540}, C_{960}, and C_{1500} with icosahedra symmetry. Iijima observed concentric shell onion-like carbon particles by transmission electron microscopy (TEM). The cage with an icosahedra shape could be constructed from sp^2 carbon atoms resulting in a bond angle of 120°. The curvature and strained surfaces are formed by sp^3 hybridized carbon atoms. A perfectly flat grapheme sheet of any size would convert into a closed cage with minimal number of defects when there are 12 pentagonal declinations. Round particles were created under pressurized helium. It dissolved in benzene.

Nanostructures can be created by pyrolytic methods and nanotube growth mechanisms have been identified. Nanostructure creation is governed by metal cluster catalyst. It was found that nanotubes could be obtained by electrolysis. Boron nitride nanotubes were synthesized. C_{60} has the same structure of a soccer ball.

2.2   Combustion Flame Synthesis

J. Howard from MIT, patented[3] the first-generation combustion synthesis method for fullerene production. A continuous low flow of hydrocarbon fuel is burned at low pressure in a flat flame. This was a significant advance over the carbon arc method developed at the University of Arizona. In this process, individual graphite rods are vaporized with electrical currents in low-pressure inert gas. The soot generated is collected and the fullerenes separated by chromatographic methods. Small quantities of fullerenes could be produced for purposes of laboratory research using this method. However, this process was found not scalable. Howard founded the Nano-C Corporation in 2001. The second-generation combustion synthesis method optimizes[4] the conditions for fullerene formation and the need for expensive post-processing is eliminated. A continuous high flow of hydrocarbon is burned at low pressure in a three-dimensional chamber. This is a scalable process. Manufacturing plants have been constructed in Japan and the United States that produce fullerenes at 40 metric tons per year.

Fullerenes with greater than 98 percent purity can be synthesized. Manufacturing conditions can be changed to suit the customer requirements of the materials. The projected price of the fullerenes has decreased to $200/kg in the second-generation process compared with $25,000/kg in the Arizona process and $16,000/kg in the first-generation process. The second-generation process is operated at pilot scale in Westwood, Massachusetts with a capacity in place scaled to several thousand kilograms per year.

Large-scale manufacture of fullerenes with higher yield can be affected using a reaction chamber with a primary zone to carry out the initial phase of the combustion synthesis (Fig. 2.2) and a secondary

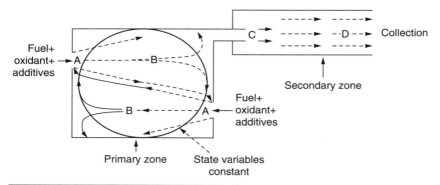

Figure 2.2   Primary zone reactor in conjunction with secondary zone reactor during combustion flame synthesis of fullerenes.

zone where combustion products with higher exit age distribution do not mix with those of lower exit age distribution. The primary zone is designed with recycle to allow for back mixing and reaction of reactants with higher exit age with fuel or oxidant and/or combustion products with lower exit age distribution. The primary zone is operated in conjunction with the secondary zone. The fuel contains carbon and is allowed to react with the oxidant to form the first combustion products such as the oxides of carbon and hydrogen. The combustion products and unreacted residue mixed with incoming carbon containing fuel and oxidant are introduced into a secondary zone under conditions favorable for fullerene production. Fuel such as unsaturated hydrocarbons, oxidants, and additives are added at optimal locations in the secondary zone for fullerene production. A combustor is provided with a chamber with multiple injection inlet ports that are positioned such that the material with smaller age of the second feed mixes with the material with larger age of the first feed. Flame control and flame stability is critical in achieving higher throughputs of fuel and oxidants. Additives such as manganese or barium compounds are effective in soot suppression. Cyclic aromatic acid anhydrides are used as additives to promote the formation of fullerene precursors such as five-membered rings in polycyclic aromatic structures. Metal additives such as cobalt, iron, and nickel have been found to enhance the formation of fullerene products. It is important to control the concentrations of radical intermediates as they are suspected to affect the formation and destruction of fullerene intermediates and products. The fullerene formation reactions are allowed for longer residence times in the secondary zone—an increase from 100 to 300 ms allowed in conventional systems to 1 to 10 s. This is achieved by use of insulation, heating or cooling of the secondary zone. The preferred range of operating parameters are a residence time in the primary zone of 2 to 500 ms, residence time in the secondary zone from 5 ms to 10 s, total overall equivalence ratio in the range of 1.8 to 4.0, pressure in the range of 10 to 400 torr, and temperatures in the range of 1500 to 2500 K. Benzene is the preferred fullerene forming fuel. A parametric study was conducted at difference equivalence rations, pressures, residence times, fuels, methods of introduction of feeds, collection methods, etc.

2.3 Crystal Formation

Fullerene crystals with various shapes can be produced at a high yield. They can be used to study fullerene crystals, members of minute machines, materials of unwoven fabrics, etc. They can form a shaft of a micromachine. Aggregates of fullerene molecules arranged in a Face Centered Cubic, FCC, lattice can possess superconductive property.

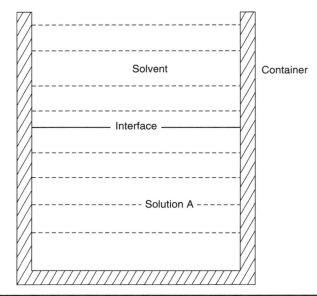

Figure 2.3   Counter diffusion and fullerene crystal formation at the interface.

A solution of fullerene with high purity in trimethylbenzene was put into a cylindrical glass container and brought in contact with isopropyl alcohol. The diameter of the cylinder was 1.9 cm.[5] The two layers of liquids (Fig. 2.3) were 6 mm deep of a lower layer of solution and 10 mm of an upper layer of solvent. A circular interface with 1.9 cm diameter was formed. The layers were covered with a lid and allowed to stand still at 20°C for 48 h. By counter diffusion from solution to the pure solvent, a number of fullerene single crystal fibers with a needle shape were formed. The solvent usually has a lower solubility of fullerene compared with the solution in the bottom layer. The needle diameters were 2 to 100 μm. These were allowed to grow in solution prior to precipitation.

Crystals of diameter in the range of 200 nm to 2 μm at a length of 0.15 to 5 mm were obtained when the fullerenes were dissolved in toluene (Fig. 2.4).

By mixing the poorer solvent B in the top layer of the cylindrical tube with the solution of fullerene, a state of supersaturation is reached. Mixing is affected by molecular diffusion and not by mechanical stirring. Crystals generated may be treated at high pressure (3 to 9 Pa), temperature of 250 to 400°C, and laser beam irradiation.

This posttreatment may be expected to result in improved purity of the fullerene crystals by removal of residual solvent remaining in the crystal. The sublimation temperature of fullerenes is 600°C. They decompose at 450°C. Ions can be positioned between fullerene molecules. This can change the properties of fullerenes a great deal. Fibrous

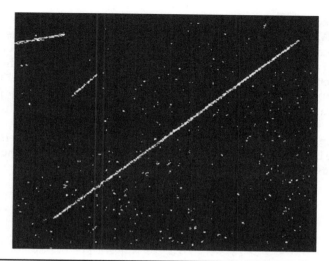

FIGURE 2.4 Optical micrograph of needle fullerene crystal structure.

crystals, flaky crystals, and particulate crystals can be formed. The reactions with other substances can be expected to occur at the surface of fullerenes due to the presence of Pi bonds. Fullerene crystals are formed into a sheet and nonwoven fabrics are made out of them. The length/diameter, L/D ratio of flaky crystals is found to be greater than 10.

2.4 Sintering

With hardness as high as that of a diamond, fullerenes can be produced using powder metallurgy methods. These are affected at high temperature and high pressure. The pressure of compaction is from 1.0 to 10.0 GPa. The temperature of the process is 300 to 1000°C and compaction time is 1 to 10,000 s. Carbon soots with diameters of particles of 0.7 to 7.0 nm were found.[6] Purity of carbon was 99.9 percent. Fullerene powder is compacted into a super hard material. These materials are nearly isotropic and polymeric. Buckyball-based sintered carbon materials can be transformed into polycrystalline diamond at temperatures and pressures less than those needed to convert graphite can. Fullerenes possess the property of super plasticity in the temperature range of 200 to 400°C and with pressures of 0.01 to 1.0 GPa.

2.5 Organic Synthesis Method

According to the isolated pentagon rule, one isomer is stable for C_{60}. Two stable isomers are possible for C_{70}, 5 for C_{78}, 24 for C_{84}, 450 for C_{100}, and so on. One hundred plus different fullerenes that obey the isolated pentagon rule can be generated from 100 carbon atoms or less. With larger fullerenes, the number of possibilities becomes more than one million. The C_{60} and C_{70} can be harvested from the soot generated from vaporization of graphite rods. In this section, the method of chemical synthesis of C_{60} in isolable quantities is provided. There are 90 C-C sigma bonds in C_{60}. This is larger compared with those present in natural products such as taxol. The strategy for synthesis of C_{60} is to construct the complete carbon cage with sp^3 hybridized CH units at large numbers of vertices followed by dehydrogenation of the ball in the last step. Barth and Lawton reported the synthesis of corannulene[7] as follows:

Sumanene was synthesized in a similar fashion. A 60-carbon cage was built[8] with 32 rings. Alternate strategies reported in the literature include assembling a 60-carbon cage with fewer than 32 rings and then collapsing it to a "soccer ball" structure through a series of transannular bond-forming reactions and dehydrogenations. Another strategy is to assemble two hemispherical hydrocarbons and then stitch the rims of the two hemispheres together. Polycyclic aromatic hydrocarbon that is threefold symmetric can be achieved by retro synthetic analysis that splits open the ball and peels back the sides. This can serve as an intermediate. The challenge is how to stitch up the seams between the arms to make the ball. The arms need to be bent and provided with the strain energy of the final target. Advances in modern organic chemistry have made synthesis of complex natural products possible. The focus these days is on specific structural features.

Corannulene ($C_{20}H_{10}$) was selected to develop a method to transform oligoarenes into highly strained curved Pi surfaces. A planar fluoranthene with two carbon substituents attached at 7 and 10 positions contains 20 carbon atoms and is a good precursor to corannulene. The carbon atoms at the edge of the rings in fluoranthene can be joined together by making use of the normal put-of-plane bending of the nucleus of the molecule. Molecular vibrations are realized at room temperature, which can distort the molecule out-of-plane, similar to closing of the book, with all 5 C atoms of the central ring pyramidzed in the same direction shortly. The amplitude of these vibrations can be increased at higher temperature to a point of bond formation and increase the degree of pyramidization. This changes the molecule in a plane to a curved bowl shape. Large amounts of thermal energy can be introduced into the molecule by flash vacuum pyrolysis (FVP). The energy would result in bending of the molecule enough to effect ring closure on a "soccer ball" structure at 1000°C. Bond formation requires proximity of atoms as well as reactivity to "attack" each other. Carbenes can be made at high temperatures by what is referred to in the literature as "Roger Brown rearrangement." Scott started with compounds available commercially and prepared 7, 10-diethynyl fluoranthene in simple steps. It was subjected to vacuum and sublimed in a quartz tube and the product was pure corannulene. Thus, it was demonstrated that heat could be used to "bend" molecules momentarily into highly distorted conformations such as bowl-shaped polyarenes. In fact, it is an oligoarene. These bonds are formed at high temperature in the gas phase.

Formation of many C-C aryl bonds was needed to stitch up the C_{60} precursor. Radical cyclizations can lead to geodesic polyarenes. Radical generation is necessary but not sufficient to achieve the desired cyclization reactions. They tried to synthesize dibenzocorannulene photochemically.

The 60-carbon ring system can be built by acid-catalyzed aldol trimerization of ketone. The sickle shaped arms of the product are made to be oriented in the same sense by the head-to-tail nature of the assembling fused benzene rings. Irradiation with UV laser of oligoarene results in conversion to C_{60} with a soccer ball structure. This can be detected by mass spectrometry. $^{13}C_{NMR}$ labeling experiments were used to show that oligoarene zips up to the "soccer ball" structure. This is affected by cyclodehydrogenations.

2.6 Supercritical Oligomerization

The graphite arc method where the soot from the graphite rods are treated with chromatographic separation methods to produce fullerenes, the combustion synthesis method, and the organic synthesis method described in the preceding sections result in low yield of fullerenes. A method was patented[9] where naphthalene was brought in contact with supercritical ethanol and ferric chloride as a catalyst for 6 h. The reaction products were subjected to extraction with toluene. Fullerenes were isolated, purified, and analyzed by laser desorption time-of-flight mass spectrum analyzer. The reactor temperature range was 31 to 500°C and the pressure ranged from 3.8 to 60 MPa. The method is lower in cost and conducted on less expensive apparatus. The reaction products from contact of supercritical fluid with aromatic precursor are subjected to a heat treatment at a temperature range of 300 to 600°C. This serves as a purification step. During heat treatment, ultrasonic waves can be used to improve the solvation of fullerenes in the solvent. The isolation of fullerenes can also be effected by filtration and centrifugation methods. The reactor vessel is shown

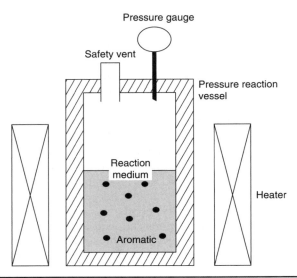

Figure 2.5   Supercritical oligomerization.

in Fig. 2.5. A safety vent is provided as higher pressures are used. Heating elements are used to provide the thermal energy. The reacting medium consists of aromatics and the supercritical fluid.

2.7 Solar Process

Smalley,[10] who shared the Nobel Prize in 1996 with Kroto and Curl for their discovery research on the third allotrope of carbon, fullerenes, patented a process to make fullerenes by tapping into solar energy. The carbon is vaporized by applying focus of solar rays and conducting the carbon vapor to a dark zone for fullerene growth and annealing. The solar energy also is used to prevent cluster formation.

They demonstrated that fullerenes could in fact be manufactured using direct sunlight as shown in Fig. 2.6. The collection of sunlight is effected by a parabolic mirror made out of electroformed nickel with rhodium finish with a focal length of 6 cm, an outer diameter of 35.6 cm, and focused on the tip of a 0.4-mm diameter graphite rod. The graphite rod as shown in Fig. 2.6 is mounted inside a 58 mm id, 2 mm wall thickness, and 20 cm in length Pyrex tube. They are arranged in a manner that it can be translated along the optic axis of the paraboloid. Heat loss minimization is achieved by enclosure of the graphite rod with helium tungsten heater with eight turns on a 3-mm diameter with a 10-mm long cylinder and with tungsten with a diameter of 0.25 mm. This can also help in annealing the carbon clusters as they grow from the graphite rod. The system was evacuated to less than 20 mtorr. Absorbed gases in the graphite rod were degassed with the tungsten heater in operation for several hours. The system was

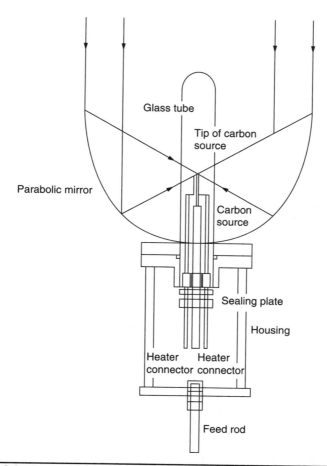

FIGURE 2.6 Apparatus for the solar process for manufacture of fullerenes.

purged and refilled with 50 torr argon and sealed off. The apparatus was mounted on the yoke of an 8 in equatorial telescope mount and adjusted so that the sunlight was focused directly on the tip of the graphite target. With careful alignment of the equatorial axis to the Earth's rotation, the motorized telescope mount was readily able to track the sun for several hours without need for more adjustment.

The fullerene generation apparatus (Fig. 2.6) was operated for 3 h on a day when the direct solar flux at the test site at an elevation of 1400 m was measured at approximately 850 w/m². The central axis of the evaporator was at an angle of 10 to 25° from the vertical. As a result, argon was heated by the tungsten heater and carried efficiently by convection upward over the solar irradiated carbon tip. The condensing vapor was quickly swept out of the intense sunlight, cooled in the upper regions of the Pyrex tube, and subsequently deposited on the upper walls.

The fullerene content of the soot deposits collected on the inside of the Pyrex tube was analyzed by extraction with toluene. Fullerenes were detected by high pressure liquid chromatography (HPLC).

The solar process obviates the heating of the plasma arc at the core in excess of 10,000°C in the electric arc process. The rate of photochemical reaction of the carbon species in the excited triplet state increases linearly with the photon flux and carbon radical concentration. In this process, concentrated solar energy is used to produce the carbon vapor. The carbon vapor is then transferred to a fullerene growth and annealing zone where the carbon in the carbon vapor reacts to form fullerenes. The vapor is then transferred to a condensing zone where the vapor condenses into soot comprising fullerenes. The growth and annealing of fullerenes proceeds more efficiently in a carbon vapor that has a relatively low concentration of carbon and produces soot with a higher yield of fullerenes. A bright zone in addition to the vaporization zone prevents the formation of carbon clusters having more than 30 carbon atoms in the carbon vapor and then passing the carbon vapor into the fullerene growth and annealing zone. The fullerene growth and annealing is maintained in relative darkness.

Metallofullerenes can be generated by contacting the carbon vapor with a suitable metal. The fullerene molecules along with graphite carbon molecules are then condensed and collected as solid soot. The fullerenes are purified by sublimation from the soot with an appropriate solvent such as benzene, toluene, or xylene followed by evaporation of the solvent to yield fullerene molecules in solid form. Doped fullerenes can be used as either p type or n type dopants in fullerene semiconductor devices. Metallofullerenes are used for making metal-carbon composites. They are used as catalysts and preparation of metal-containing polymers and glass. Ink compositions use fullerenes. They can also be used in research for room temperature superconductors.

2.8   Electric Arc Process

The objectives of the development of the electric arc process for fullerene production are to increase the yield and hence the production rates and make the process scalable to large plants so that fullerenes can be manufactured in commercial quantities. Carbon material is heated using an electric arc between two electrodes to form a carbon vapor. Fullerene molecules and graphite carbon molecules are condensed later and collected as soot material. Fullerenes are later purified by extraction of the soot using a suitable solvent followed by evaporation of the solvent to yield the solid fullerene molecules.

A cross-sectional view of a carbon arc fullerene generator patented by Nobel laureate R.E. Smalley[11] is shown in Fig. 2.7. It comprises a vaporization chamber inside an enclosure body made out of material that is capable of withstanding the temperature and pressure

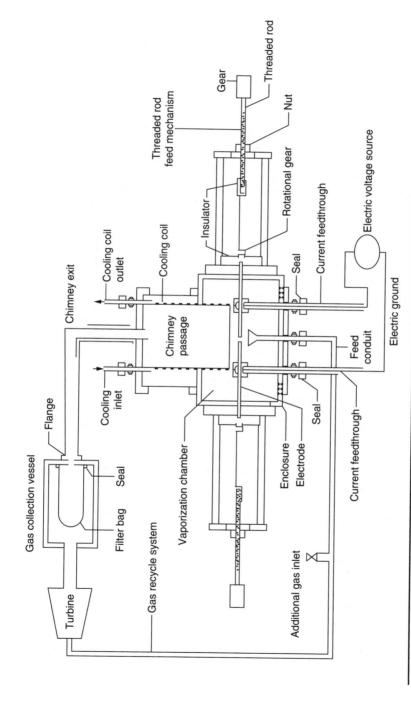

Figure 2.7 Electric arc process for fullerene production.

35

required. One example is stainless steel. Electrodes are placed within the vaporization chamber. The electrodes are connected to an electric voltage source via an electrical conductor, which passes through a water-cooled current feed-through. The current feed-through passes through the walls of the enclosure body and is insulated from the electrical conductor. Seals are provided to prevent outside atmosphere from entering the vaporization chamber. The electrode and electrical current source are connected by an electrical conductor. Cooling is provided to prevent the conductor from melting and to reduce power loss. The electrodes are capable of being rotated and the electric contact is made by gimble wheel loaded rod contact. The electrode is shielded from the enclosure body using an insulator. Sometimes the vaporization chamber may be operated below atmospheric pressure. Rotation gear is attached concentrically to the electrode. The rotational drive mechanism allows for rotation of the gear around the longitudinal axis, by use of a continuous drive belt around rotation gear and around a motorized gear that, when rotated, pulls the continuous drive belt around the rotation gear resulting in spinning motion of the rotation gear and electrode. As the electrodes are consumed during the fullerene production, some means for advancing the electrodes toward the gap area is provided. The desired gap for the electric gap also needs to be borne in mind. The threaded rod feed mechanism comprises a threaded rod fixed to gear at one end. The threaded rod passes through the threaded nut and engages the insulator. Rotational motion of the electrode is prevented from being transferred to the threaded rod by the insulator. It also electrically isolates the electrode from the threaded rod. Further, it serves as a movable platform that determines the position of the electrode. Insulator, electrode, and threaded rod are cooperatively coupled so that the electrode may be pushed or pulled by rotation of the threaded rod. The electrode is grounded outside the vaporization chamber. Electric ground is radioactively heated by the arc and some cooling is provided. The speed of rotation of the electrodes ranges from 1 to 100 rpm. Feed conduit into the feed chamber and a chimney passage are provided in the fullerene generator. Cooling coils allow for absorption of heat from the chimney passage. The warmed fluid may be disposed or cooled and recirculated back to the cooling coil inlet.

The chimney exits lead into the gas recycle system. A gas collection vessel, a filter bag secured with the inside of the flange, forms the gas recycle system. The solid particles remain in the filter bag and the clean gas passes through the walls of the filter bag into the remaining portion of the gas collection vessel. The vaporization chamber needs to be dehydrated. The vaporization chamber is evacuated to a pressure less than 0.01 torr. The vapor containing the fullerene soot flows to the chimney passage and condenses at low temperatures. Soot that does not condense on the walls is collected in the filter bag. A cyclone separator may be used to separate the solid particles from the clean gas. The process may be continued until the electrodes have been consumed, at which point the electrical voltage may be withdrawn. Carbon soot lining the walls of the chimney passage and filter bag may then be recovered. A selective solvent that can dissolve fullerene and not the graphite can be used to extract the fullerene compounds. Evaporation of the solvent from the extract will allow for recovery of fullerenes as the solid residue. Separation of fullerenes from graphite may also be achieved by boiling solvents, ultrasonic sonification separations, supercritical fluid extraction, sox let extraction, and other methods. After carbon condensation stops, carbon soot may be recovered by opening the system by removing the chimney sap and gently scrapping or brushing the carbon soot from the chimney walls.

A separate zone is provided where the growth and formation of fullerenes are promoted. This is the fullerene annealing zone. The atmosphere is one where the temperature, pressure, and residence time favor the growth of fullerenes. The range of temperatures of operation of the annealing zone is 1000 to 2000°C. The optimal residence time ranges from 1 msec to 1 sec. The fullerenes formed by the electric arc are carried away by inert gas into a fullerene annealing zone above the electric arc but below the chimney passage.

2.9 Applications

2.9.1 Superconductors

Fullerenes have a soccer ball shape molecular architecture. The electrons in the π orbital are delocalized and can be tapped into when synthesizing superconductors.

It was found that alkali-doped fullerenes are known to exhibit superconducting properties. The transition temperature of alkali bases C_{60} fullerene molecules is low at approximately 40 K. Hence, it is not suitable for electrical circuit material that requires fine fabrication. NEC patented a method[12] where C_{20} fullerene molecules were polymerized into a single dimensional chain and bound in the sp^3 hybridization form. C_{20} fullerene molecules possess stronger electron-lattice interaction compared with that of C_{60} fullerene. These polymers were found to have a transition temperature of 100 to 150 K (Fig. 2.8). The chemical stability of C_{20} polyfullerenes is higher compared with that of oxide superconductor. Thus, materials with higher superconductivity and better chemical stability can be obtained.

2.9.2 Adsorbents

Fullerenes can be used in the polymeric form to make adsorbents with improved gas storage capability. Gas-solid adsorption has been used in the past 20 years to develop gas storage technologies. Lack of

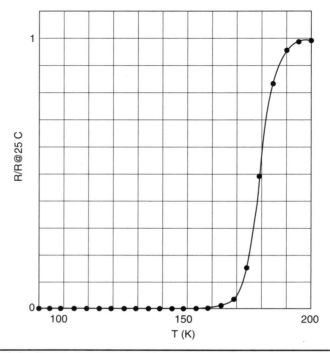

FIGURE 2.8   Electrical resistance of 1-D C_{20} chain with temperature.

high performance adsorbent materials has delayed the commercialization of adsorption-based gas storage technology. Materials suitable for gas storage applications are characterized by micropores with diameter less than 2 mm and mesopores with diameter range of 2 to 50 mm. With application of pressure, more gas molecules will be adsorbed and when gas pressure is reduced the adsorbed molecules will leave the pore surface. Gas adsorption and desorption are reversible processes. The gas storage performance of an adsorbent can be evaluated using the two criteria of equilibrium adsorption capacity and dynamic adsorption/desorption properties. The equilibrium adsorption capacities are characterized by gravimetric adsorption and volumetric adsorption capacity. The dynamic adsorption property is the dynamic adsorption/desorption rate, adsorption/desorption recyclability, and adsorption/desorption hysteresis. High surface area activated carbons and zeolites are the most researched gas adsorbents. The surface areas of activated carbon range from 100 to 4000 m^2/gm. Activated carbons are made from carbonaceous materials such as coal pitch, coconut shells, and petroleum wastes. The bulk densities of activated carbons are in the range of 0.1 to 0.7 gm/cm^3. Zeolites are porous crystalline aluminosilicates. The zeolite framework consists of assemblage of SiO_4 and AlO_4 tetrahedral molecular structures fused together in various arrangements through shared oxygen atoms forming crystalline lattice containing pores of nanometer dimensions into which gas molecules are adsorbed. Bulk density of zeolites are higher, ranging from 0.5 to 1.5 gm/cm³. Gravimetric adsorption capacities of zeolites are low and adsorption/desorption capacities of zeolites are not as high as are those of activated carbon. They exhibit large desorption hysteresis.

Fullerene-based adsorbents have a bulk density of at least 1.4 gm/cm³. Fullerene-based material is made of fullerene that is polymerized at high temperature under inert gas pressure allowing the formation of a cage structure. For example, 1,4-phenylenediamene (PDA) is used to form fullerene-PDA polymer structures and hexamethylenediamine (HMDA) are used to form fullerene-HMDA polymer structures. These polymers are activated to increase their gas adsorption capacity by incorporation of additional micropores. The temperature range of a post-polymerization activation process is 350 to 850°C.

Adsorbents[13] made from polymeric fullerenes possess higher gravimetric gas adsorption capacities compared with the top of the line activated carbon and excellent gas adsorption/desorption properties. The unique properties of fullerenes are pelletizability of fullerenes, resulting in higher volumetric gas storage capacity, and the capability of adsorbing gas molecules in large interstitial spaces inherent in the face centered cubic (FCC) crystal structure. Fullerenes crystallize into an FCC structure. As the effective molecular diameter of C_{60} is approximately 1 nm, the interstices are big enough to accommodate gas molecules. Although the kinetic volume of gas molecules is less than 0.35 nm, the octahedral sites have a volume larger than 0.42 nm³. The gas molecules can be attracted by the interstices by van der Waals forces. The interstices serve as micropores. When fullerenes are polymerized, the cage structure is opened and the inside closed cage spaces are also available for gas molecules. The unique structure of polyfullerenes results in high gas storage capacities. The polyfullerenes are denatured and are no longer soluble in toluene. Micropores in polyfullerenes are less than 1 nm in diameter. Extra micropores can be generated by chemical modification after fullerene polymerization. Oxygen storage units can be designed using polyfullerenes.

2.9.3 Catalysts

Fullerenes can make excellent catalysts. Technologists at Fritz Haber Institute in Berlin, Germany, have used Bucky-onions to convert ethyl benzene into styrene. The yield of styrene was 62 percent. This was an improvement from the 50 percent maximum yield in existing processes. Conversion of ethylbenzene to styrene is one of the top 10 industrial chemical processes. SRI International has developed catalysts using fullerenes for more efficient hydrogenation/dehydrogenation of aromatics, for upgrading heavy oils, and for conversion of methane into higher hydrocarbons by pyrolytic or reforming processes.

2.9.4 Composites

Fullerene composites with improved mechanical strength[14] in addition to inherent properties of the fullerenes as plastic deformation and work hardening can be prepared. The C_{60} fullerenes formed the matrix and the CNTs are reinforcing materials. The matrix is made out of ultrafine fullerene particles of 5 to 50 nm in diameter and the reinforcing material is a mixture of nanotubes, nanocapsules, and some impurities (Fig. 2.9). The volume fraction of the reinforcing materials are 15 to 45 percent by weight. The fullerene composites can be used as superconducting materials, semiconducting materials, catalysts, and materials for nonlinear optics.

2.9.5 Electrochemical Systems

Fullerenes can be used to develop advanced electrochemical systems. A proton-conducting electrode with a three-phase interface is used in the construction of the electrochemical device.[15] The proton-conducting electrode contains a mixture of a fullerene derivative, electron conducting catalysts, and a proton dissociating group. A proton dissociating group is a functional group capable of releasing protons on electrolytic dissociation. The electrode has enhanced properties. The fullerene derivative used to prepare the electrode exhibits negligible atmosphere dependency and allows for enhanced proton conductivity

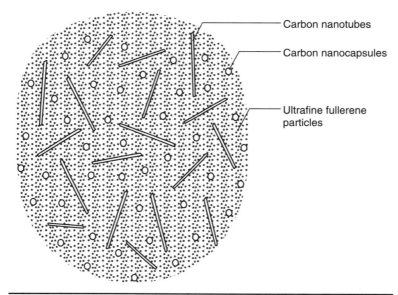

Carbon nanotubes

Carbon nanocapsules

Ultrafine fullerene particles

Figure 2.9 Fullerene composite.

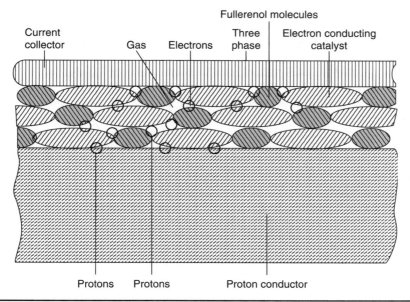

Figure 2.10 Proton conductor and electrochemical systems.

even in a dry atmosphere. A coating of a mixture of fullerene derivative and proton dissociating group is applied on a gas transmitting current collector. The mixture can be coated in multiple layers and the desired film thickness can be produced.

The electrochemical device is made by sandwiching a proton conductor between two electrodes. The electrodes are made porous so that the gas can be permeated throughout the entire electrode. A three-phase interface is produced not only in the vicinity of a contact point between the proton conductor and the catalyst, but also in the vicinity of a contact point between the catalyst and the fullerenol molecules (Fig. 2.10). The three-phase interface is a site where the electrons, protons, and gases all meet simultaneously. The electrode can be a gas electrode. A fuel battery employs the proton-conducting electrode. The mechanism of proton conduction of a fuel cell consists of dissociation of protons and migration of the dissociated protons from a hydrogen electrode to an oxygen electrode.

2.9.6 Synthetic Diamonds

There is a way to convert fullerenes into synthetic diamonds. The cost of conversion is not favorable for large-scale production. However, this can change when the cost of production of fullerenes comes down with advances in technology. Diamonds are the hardest known substance in the universe. They possess the highest thermal conductivity of any known substance in the universe. The room temperature thermal conductivity of diamond is 5 times the thermal conductivity of copper. As a result, diamonds are used for cutting tools. General Electric synthesized diamonds for the first time in 1955. The methods used require high pressures and temperatures and include static crystallization, molten metals or alloys, shock conversion from graphite, and static conversion.

A method was developed to produce diamond crystals from a metallofullerite matrix independent of external application of pressure.[16] A metal-carbon matrix of an allotropic metal and metallofullerites of the allotropic metal are heated together up to their critical temperature and then hyper-quenched rapidly to collapse the fullerene structures in the matrix into diamond crystals. Hyper-quenching involves speedy cooling that causes a phase matrix change. This is achieved by immersion in a cold fluid such as water or oil.

2.10   Summary

Fullerenes (C_{60}) are the third allotropic form of carbon. The Nobel Prize for their discovery was awarded in 1996 to Curl, Smalley, and Kroto. Soccer ball structured, C_{60}, with a surface filled with hexagons and pentagons, satisfies Euler's law. Euler's law states that no sheet of hexagons will close. Pentagons have to be introduced for hexagon sheets to close. Stability of C_{60} requires Euler's 12-pentagon closure principle and the chemical stability conferred by pentagon nonadjacency. C_{240}, C_{540}, C_{960}, and C_{1500} can be built with icosahedra symmetry.

Howard patented the first-generation combustion synthesis method for fullerene production, an advance over the carbon arc method. The second-generation combustion synthesis method optimizes the conditions for fullerene formation. A continuous high flow of hydrocarbon is burned at low pressure in a three-dimensional chamber. Manufacturing plants have been constructed in Japan and the United States with production capacity of fullerenes at 40 metric tons per year. Purity levels are greater than 98 percent. The reaction chamber consists of a primary zone where the initial phase of combustion synthesis is conducted and a secondary zone where combustion products with higher exit age distribution do not mix with those of lower exit age distribution. Flame control and flame stability are critical in achieving higher throughputs of fullerenes. Typical operating parameters include residence time in the primary zone of 2 to 500 ms, residence time in the secondary zone from 5 ms to 10 s, total equivalence ratio in the range of 1.8 to 4.0, pressure in the range of 10 to 400 torr, and temperature in the range of 1500 to 2500 K.

Fullerene crystals can be produced at high yield. By counter diffusion from fullerene solution to pure isopropyl alcohol solvent, fullerene single crystal fibers with needle shape were formed. Needle diameters were found to be 2 to 100 μm and lengths were 0.15 to 5 mm.

Buckyball-based sintered carbon materials can be transformed into polycrystalline diamonds at less severe conditions using powder metallurgy methods.

A chemical route was developed by Scott to synthesize C_{60}. Corannulene is synthesized from a naphthalene structure. As the rings fuse and the sheet forms, it is rolled into a soccer ball structure. The challenge is how to stitch up the seams between the arms to make the ball. Oligoarenes are transformed into highly strained curved Pi surfaces. The molecule needs to bend to effect ring closure on a soccer ball structure at 1000°C. A 60-carbon ring system can be built by acid catalyzed aldol trimerization of ketone. Oligoarene zips up to the soccer ball structure affected by cyclodehydrogenations.

In order to generate higher yield, supercritical ethanol was used to react with naphthalene with ferric chloride as a catalyst for 6 h. The reaction products were subjected to extraction with toluene. The reactor temperature range was from 31 to 500°C and the pressure range was from 3.8 to 60 MPa. Smalley patented a process to make fullerenes by tapping into solar energy. The carbon is vaporized by applying focus of solar arrays and conducting the carbon vapor to a dark zone for fullerene growth and annealing. Fullerene content of soot deposits collected on the inside of the Pyrex tube was analyzed by extraction with toluene.

In the electric arc process for fullerene production, carbon material is heated using an electric arc between two electrodes to form carbon vapor. Fullerene molecules are condensed later and collected as soot. Fullerenes are later purified by extraction of soot using a suitable solvent followed by evaporation of the solvent to yield the solid fullerene molecules.

Applications of fullerenes include: higher temperature superconductors; polymerized fullerene molecules with transition temperatures of 100 to 150 K; adsorbents with improved gas storage capability; excellent catalysts such as Bucky onions to convert ethylbenzene to styrene, for example; fullerene composite with improved mechanical strength; advanced electromechanical systems where a proton conductor is sandwiched between two porous electrodes; synthetic diamonds by static crystallization; and shock conversion methods.

Review Questions

1. What is Euler's Theorem?

2. According to Euler's Theorem, is C_{540} a feasible fullerene structure?

3. What is the molecular principle underlying rotational spectroscopy?

4. Why does a $^{13}C_{NMR}$ spectrum of C_{60} fullerene show a singular spectrum?

5. What are the four known allotropes of carbon?

6. Give an example of quantum angular momentum calculation used by Kroto to show the stability of C_{60} fullerene structure.

7. How were radio telescopes used during the discovery of C_{60} fullerenes?

8. What does $5 \times 12 = 60$ mean in fullerene chemistry?

9. What are magic numbers?

10. What is the isolation principle?

11. What are endohedral fullerenes?

12. Can allotropic chromatography be used to synthesize C_{60} fullerenes?

13. Why is the carbon arc method used to produce C_{60} fullerene not good for industrial practice?

14. What are the advantages of using the combustion synthesis method?

15. How was the cost of C_{60} fullerene reduced from \$25,000/kg from the Arizona process to \$200/kg in the second-generation combustion flame synthesis method?

16. Why is the yield in the processes to manufacture C_{60} fullerene low?

17. What is the difference in the function of the primary zone and secondary zone in the combustion flame synthesis method?

18. What is the typical residence time used in the combustion flame synthesis method?

19. Where are fullerene crystals used?

20. What is the phenomena of "counter diffusion"?

21. Which liquid has higher salvation power during the process to make fullerene crystals—the one in the solution or the one in the pure solvent?

22. What is the shape of the fullerene crystal—needle or spherical? Why?

23. Why is sintering done after fullerene production?

24. Where is the isolated pentagon rule used?

25. What is the structure of corannulene?

26. How is the cage structure formed during organic synthesis of C_{60} fullerene?

27. Which bonds help in the formation of geodesic polyarenes?

28. What is the role of strain energy during the formation of the cage structure of C_{60} fullerene?

29. What happens when naphthalene is contacted with supercritical ethanol and ferric chloride catalyst?

30. How is solar energy used in the patented process for fullerene production by Smalley?

31. Is the typical operation time of fullerene generation apparatus by solar energy 3 h/day, 3 min/day, or 3 days/month?

32. What are the advantages of using solar energy for fullerene production?

33. What were the objectives for development of the electric arc process for fullerene production?

34. What is meant by a threaded gear mechanism?

35. What happens in the fullerene annealing zone?

36. Give two methods for separating fullerenes from graphite.

37. How does polyfullerene result in 100 to 150 transition temperature for superconductivity?

38. What are the two properties of good gas adsorbents?

39. What is the advantage of using Bucky onions to make styrene from ethyl benzene?

40. What forms the matrix compound of fullerene composition?

41. How is C_{60} fullerene used in an improved electrochemical system?

42. Why is the world not filled with synthetic diamonds from C_{60} fullerene as of yet?

43. What is hyper-quenching?

44. What are some interesting properties of C_{60} fullerene?

References

1. Sir Harold W. Kroto, *Symmetry, Space, Stars and C_{60}*, Nobel Lecture, http://nobel.se, December 1996.
2. S. Iijima, *Helical microtubules of graphitic carbon, Nature*, 354, 56–58, 1991.
3. J. B. Howard, D. F. Kronholm, A. J. Modestino, and H. Richter, Method for Combustion Synthesis of Fullerenes, US Patent 7,396,520, 2008 Nano-C Inc., Westford, MA.
4. J. B. Howard, J. T. McKinnon, Y. Makaraovsky, A. L. Lafleur, and M. E. Johnson, *Fullerenes C_{60} and C_{70} in flames, Letters to Nature*, 352, 139–141, 1991.
5. T. Yoshii, Fullerene Crystal and Method for Producing Same, US Patent 7,351,284, 2008, Nippon Sheet Glass Co., Tokyo, Japan.
6. O. E. Voronov and G. S. Tompa, Fullerene Based Sintered Carbon Materails, US Patent 6,783,745, 2004, Diamond Materials, Piscataway, NJ.
7. W. E. Barth and R. G. Lawton, Journal of American Chemical Society, Vol. 93, 1730, 1971.
8. L. T. Scott, *Methods for the chemical synthesis of fullerenes, Angewandte Chemie*, 43, 4994–5007, 2004.
9. S. Kawakami, T. Yamamoto, and H. Sano, Method for Producing Fullerenes, US Patent, 6,953,564, 2005, Canon Kabushiki Kaisha, Tokyo, Japan.
10. R. E. Smalley, Solar Process for Making Fullerenes, US Patent 5,556,517, 1996, Rice University, Houston, TX.

11. R. E. Smalley, Electric Arc Process for Making Fullerenes, US Patent 5,227,038, 1993, Rice University, Houston, TX.
12. Y. Miyamoto, Superconducting Material and Method for Producing the Same, US Patent 7,189,681, 2007, NEC Corporation, Tokyo, Japan.
13. R. O. Loutfy, X. C. Lu, W. Li, and M. G. Mikhael, Gas Storage Using Fullerene Based Adsorbents, US Patent 6,113,673, 2000, Materials and Electrochemical Research, Tucson, AZ.
14. S. Tanaka, Fullerene Composite, US Patent 5,648,056, 1997, Research Development Corp. of Japan, Japan.
15. K. Hinokuma and M. Imazato, Proton Conducting Electrode, Method for Preparation Thereof and Electro-Chemical Device, US Patent 7,087,340, 2006, Sony Corp., Tokyo, Japan.
16. R. C. Job, Method of Producing Diamond Crystals from Metallfullerite Matrix and Resulting Product, US Patent 5,449,491, 1995, MicroMet Technology, Monroe, NC.

CHAPTER 3

Carbon
Nanotubes (CNT)

Learning Objectives

- Discovery of CNT as elongated cousins of fullerene and stable allotrope of carbon
- Five synthesis methods for CNTS: (1) electric discharge, (2) laser ablation, (3) chemical vapor deposition (CVD), (4) high pressure carbon monoxide (HIPCO), and (5) surface mediated vertical alignment of tubes
- Familiarization of physical properties and applications of CNTs
- Morphology of CNTs

3.1 Discovery

Carbon nanotubes (CNTs) are hollow cylinders made by rotation of a graphene sheet of atoms about its needle axis (see Chap. 2, Fig. 2.1). They are one category of nanostructures. They exhibit interesting physical and thermal properties. The diameters of CNTs are in the order of magnitude of nanoscale and range from 0.7 to 100 nm. CNTs are a bit elongated and their lengths are of the order of a few microns. Nanotube positions can be manipulated and their shapes can be changed. They can be "cut" and placed on electrodes. Elongated nanotubes consist of carbon hexagons arranged in a concentric manner with both ends of the tube capped by a pentagon containing Buckminster fullerene-type structures. They make excellent thermal conductors, electrical conductors, and good semiconductors. The chirality of the arrangement of graphitic rings in the walls of the nanotube and diameter are sensitive parameters that affect the property of the semiconductors made out of CNTs. CNTs with varied sizes can be concatenated and made into nanowires. Nanowires exhibit excellent electrical, magnetic, nonlinear, optical, thermal, and mechanical properties. The mechanical behavior of nanotubes can be simulated by calculation of forces acting between nanotubes and other objects such as a substrate. CNTs have a wide range of applications. They are de novo. New products can be developed and existing products can be improved upon using the unique combination of properties exhibited by CNTs. Young's modulus of a Single Wall Nano Tube, (SWNT) has been estimated at 1 TPa and the yield strength can be as high as 120 GPa. IBM uses single and multiwalled nanotubes to make channels of field effect transistors (FETs). They have also developed methods to form nanotubes in the ring form. Most nanotubes are in the straight form.

Nanotubes can behave as metals or semiconductors by virtue of their structure. They also make strong materials and exhibit good thermal conductivity. These characteristics have generated tremendous interest in their application in the nanoelectronic and nanomechanical devices. For instance, they can be used as nanowires or as active components in electronic devices such as FET.

In 1991, Iijima verified fullerenes and observed multiwalled nanotubes formed from carbon arc discharge. Two years later, along with Bethune at IBM, he observed the SWNT. He is the discoverer of CNTs.

An atomic force microscope (AFM) can be used to change the position, shape, and orientation of CNTs. An image of a CNT is obtained by use of a scan of the tip of an AFM in the noncontact mode. An AFM tip is then brought down to the surface and is used as a tiny plow in order to move the CNT. Due to the strong interaction between the CNT and the surface via van der Waals forces, the bent CNT stays where it has been placed and maintains its shape rather than snapping back to the preferred straight configuration.

Single-walled carbon nanotubes (SWNTs) were independently discovered in early 1993 by scientists at IBM Almaden Research Center and at NEC in Japan. Both the IBM and NEC groups found that transition metals covaporized with carbon catalyze the formation of SWNT with a narrow range of diameters around 1 nm. IBM used cobalt and NEC used iron. The June 17, 1993 issue of *Nature* carried back-to-back papers describing the results of the two groups at IBM and NEC. The American Physical Society awarded the J. C. McGroddy Prize for new materials in March 2002 jointly to D. S. Bethune of IBM and S. Iijima of NEC for their independent discoveries of SWNT. The citation of the prize is as follows:

> for the discovery and development of SWNT which can behave like metals or semiconductor, can conduct electricity better than copper can transmit heat better than diamond and rank among the strongest materials known

Two years later, in July 2004, the American Carbon Society medal went to Bethune; Professor M. Endo, Shinsu University, Japan; Professor S. Iijima, Director, Research Center for Advanced Carbon Materials, Meijo University and NEC special research fellow, Japan; for their work on CNTs. The citation of the medal reads as follows:

> for outstanding contributions to the discovery of and early synthesis work on CNTs

3.2 Synthesis of CNTs

There are five main methods for the synthesis of CNTs currently used around the world. These are:

1. Arc discharge
2. Laser ablation
3. Chemical vapor deposition (CVD)
4. HIPCO process
5. Surface mediated growth of vertically aligned tubes

A summary of the main features of the CVD, laser ablation, and arc discharge processes is presented in Table 3.1.[1]

3.2.1 Electric Arc Discharge Process

The electric arc discharge process works by utilizing two graphite electrodes in an arc welding type process. The welder is turned on and the rod ends are held against each other in an argon atmosphere to produce or grow CNTs. Yield of CNTs by this process is low and growth of CNT orientation is random in nature.

An arc electrode apparatus for synthesizing CNTs is shown in Fig. 3.1 as patented by Sony, Tokyo, Japan.[2] Carbon nanostructures such as SWNT, Double Walled Nano Tube (DWNT), Multi Walled Nano Tube (MWNT), fullerene, endohedral metallofullerene, and carbon nanofiber can be synthesized using electric discharge between a cathode and an anode. Some catalyst can be mixed in the anode/cathode. The single arc limits the control on the final nanostructure formed. Generated soot that deposits on the walls of the chambers needs to be harvested. Some investigators have found health hazards while soot harvesting. Therefore, soot harvesting must be done with care especially when large chambers are used.

The patent from Sony[2] discusses greater control on the final formation of the nanostructure. The direction and region of arc plasma is allowed to be adjusted to gain control on the formation of the CNT. The synthesis apparatus consists of a chamber, first electrode, second electrodes, and an adjusting mechanism. The chamber includes walls,

Method	Arc discharge method	Chemical vapor deposition	Laser ablation (vaporization)
Invented	NEC, Japan, 1992	Nagano, Japan, 1993	Smalley, Rice, 1995
Process	Connect two graphite rods to a power supply, place them a few millimeters apart, and throw the switch. At 100 amps, carbon vaporizes and forms hot plasma.	Place substrate in oven, heat to 600°C, and slowly add a carbon-bearing gas such as methane. As gas decomposes it frees up carbon atoms, which recombine as NTs.	Blast graphite with intense laser pulses; use the laser pulses rather than electricity to generate carbon gas from which the NTs form; prodigious amounts of SWNTs.
Typical Yield	30 to 90 percent	20 to 100 percent	Up to 70 percent
SWNT	Short tubes with diameters of 0.6 to 1.4 nm.	Long tubes with diameters ranging from 0.6 to 4 nm.	Long bundles of tubes (5 to 20 nm), with diameter from 1 to 2 nm.
MWNT	Short tubes with inner diameter (1 to 3 nm) and outer diameter (10 nm).	Long tubes with diameter ranging from 10 to 240 nm.	MWNT synthesis is possible.
Pros	Can easily produce SWNT, MWNTs. SWNTs have few structural defects; open air synthesis possible.	Easiest to scale up to industrial production; long length, simple process, SWNT diameter controllable, quite pure.	Primarily SWNTs, with good diameter control and few defects. The reaction product is quite pure.
Cons	Tubes tend to be short with random sizes and directions; often needs a lot of purification.	NTs are usually MWNTs and often riddled with defects.	Costly technique because it requires expensive lasers and high power requirement, but is improving.

TABLE 3.1   A Summary of Three Major Production Methods of CNTs and their Efficiency

FIGURE 3.1
Electric arc
discharge
apparatus for
synthesis of CNT.

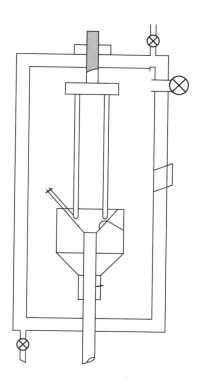

FIGURE 3.1
Electric arc discharge apparatus for synthesis of CNT.

which bound the chamber interior. The walls are structured to allow a cooling fluid to flow through. The cooling fluid enters through a cooling fluid inlet port and may exit through a cooling fluid outlet port to cool the chamber interior. An inert gas atmosphere can be produced in the chamber using suitable inlet and outlet. The inert gas mixture may be composed of hydrogen and argon. The pressure in the chamber is subatmospheric and is expected to be 300 to 760 torr.

3.2.2   Laser Ablation Process

The second method of generating CNTs is by laser ablation. In 1996, in *Science,* A. Thess, R. Lee, P. Nikolaev, H. Dai, P. Petir, J. Robert, C. Xu, Y. H. Lee, S. G. Kim, A. G. Rinzler, D. T. Colbert, G. E. Scuseria, D. Tomanek, J. E. Fischer, and R. E. Smalley wrote an article on crystalline ropes of metallic CNTs that described a laser ablation method for generating CNTs. In this process, a metal catalyst particle such as a nickel-cobalt alloy is mixed with graphite powder in a planned proportion and the mixture is pressed to obtain a pellet. A laser beam is irradiated on the pellet. The carbon and the nickel-cobalt alloy are evaporated by the laser beam. The carbon vapor is condensed in the presence of a metal catalyst. SWNTs are formed during the condensation. The early investigators found that the SWNTs were not constant in diameter.

One possible reason for this is the variation of the ratio between the carbon vapor and the metal catalyst vapor with time during absorption of laser light. The black graphite powder took more of the light compared with the metal catalyst. This can lead to uneven heating and the temperature of the graphite would increase more compared with the metal catalyst. The metal catalyst can be expected to be left on the surface. This can lead to variation in the diameter of the CNTs. Purity is not uniform. A patent by Iijima[3] of NEC, Japan, describes a process to produce SWNTs of uniform diameter. Here the carbon and metal catalyst are evaporated independently.

The laser ablation apparatus suggested is shown in Fig. 3.2. The process starts with the preparation of a carbon pellet, a metal catalyst pellet, and a laser ablation system. The carbon pellet is formed from graphite. The graphite consists of carbon and is shaped into the carbon pellet by a standard palletizing machine. The carbon pellet has a plate-like configuration and is 10 mm long and 3 to 5 mm wide. The metal catalyst pellet is formed of a nickel-cobalt alloy in the atomic ratio of 1:1. The metal catalyst is also shaped into a plate-like configuration by using the palletizing machine. The laser ablation system includes a reactor, evacuation system, inert gas-supply subsystem, heater to control the temperature of the reactor, etc. The reactor is made of a cylindrical construct and is quartz or ceramic.

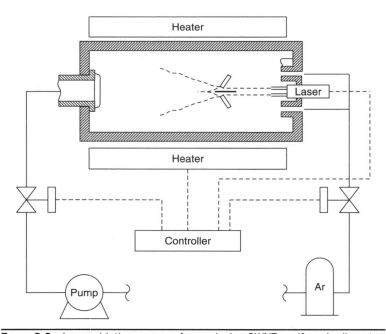

Figure 3.2   Laser ablation process for producing SWNTs uniform in diameter.

The carbon pellet and the metal catalyst pellet are separately provided inside the reactor and the spacer of the quartz plate. The spacer is 300 nm in thickness. The laser beam is shined on the carbon pellet and metal catalyst pellet, respectively. The condensate is captured using a collector. The controller is built in to control the process. The laser beam generator includes a laser light-emitting element formed of neodymium, (Nd) contains single crystalline yitrium, aluminum garnet. The laser light-emitting element radiates laser pulses. The laser light has a 532-nm wavelength and oscillates at 10 Hz. The pulse width ranges from 7 to 10 nsec. The power is regulated to 1.2 to 9.2 J/pulse. The laser cross-sectional area is 0.2 cm². The heater heats the reactor and the temperature of the reactor is controlled at 1200°C using the controller.

When the carbon pellet, metal catalyst pellet, and laser ablation system are prepared, an operator inserts the carbon pellet and the metal catalyst pellet into the reactor. The carbon pellet and the metal catalyst pellet symmetrically decline with respect to the center line of the reactor and the major surface of the carbon pellet is opposed through the spacer to the concave surface of the metal catalyst pellet. The reactor is closed and the rotary vacuum pump evacuates the air from the reactor. When the vacuum is developed in the reactor, the inert gas supply system supplies the argon gas at 0.5 l/min. The inside of the reactor is maintained at 600 mmHg. The argon gas flows from the nozzle toward the collector. Subsequently, the laser beam generator radiates the laser beams to the carbon pellet and the metal catalyst pellet. In this instance, the pulse width and the power are adjusted to 10 nsec and 50 mJ/pulse.cm². The laser beams directly heat the carbon pellet and the metal catalyst pellet, and the carbon vapor/cluster and nickel-cobalt vapor/cluster are constantly generated from the carbon pellet and the metal catalyst pellet, respectively. The carbon vapor/cluster and the nickel-cobalt vapor/cluster are carried by the argon gas toward the collector. The carbon vapor/cluster is mixed with the nickel-cobalt vapor/cluster and forms into SWNTs. The SWNTs are carried toward the collector and are captured by the collector. The carbon vapor/cluster and the nickel-cobalt vapor/cluster are constant in mass and keep the content of carbon in the condensate or the SWNT constant. This results in the constant diameter of the SWNTs. Thus, in this process the carbon and metal are separately vaporized. The carbon and metal can be in powder form and need not be in pellet form alone.

3.2.3 CVD

CVD and related techniques such as plasma-enhanced CVD (PECVD) can be used to grow CNTs. CVD is amenable for nanotube growth on patterned surfaces suitable for fabrication of electronic devices, sensors, field emitters, and other applications where controlled growth over masked areas is needed for further processing.

A variety of CNT structures is possible by CVD and related techniques. A SWNT can be viewed as a rolled-up tubular shell of graphene sheet, which is composed of benzene-type hexagonal rings of carbon atoms. MWNT can be considered as a stack of graphene sheets rolled-up into concentric cylinders. The walls of each layer of the MWNT are parallel to the central axis. Stacked cone arrangements, also known as Chevron structures, are possible. It can be an ice-cream cone structure or a piled-cone structure. As markets for CNTs are not as developed as the semiconductor equipment industry, the equipment to prepare CNTs is largely homemade, low-throughput batch reactors.

The thermal CVD apparatus for CNT growth consists of a quartz tube of diameter of 1 to 2 in inserted into a tubular furnace capable of temperature control to within ±1°C over a 25-cm zone. It is atmospheric pressure CVD with a hot open up wall system. The growth is catalyst promoted at temperatures of 500 to 1000°C. Either a cold wall or hot wall system can be effective for CNT growth. Substrate is smaller than 1 in and is placed inside the quartz tube. In the thermal CVD process, either carbon monoxide or some hydrocarbon such as methane, ethane, ethylene, acetylene, or other higher hydrocarbons is used without any dilution. The feedstock is metered through a mass flow controller.

During a typical growth run, the reactor is purged with argon or some other inert gas until the reactor reaches the desired growth temperature. Then the gas flow is switched to the feedstock for the specified growth period. At the end the gas flow is switched back to the inert gas while the reactor cools down to 300°C or lower before exposing the nanotubes to air. Typical growth rates range from a few nanometers per minute to 2 to 5 μm/min.[4] CNT growth is largely empirical. There are very little modeling studies undertaken. The effects of reactor length and diameter, flow rate, etc. on the growth characteristics are completely unknown. For growth on substrates, the catalyst mixture needs to be applied to the substrate prior to loading it inside the reactor. This is the supported catalyst approach. In the floating catalyst approach, large quantities of CNTs can be grown. A nozzle system may be used to inject vaporized catalyst precursor into the flowing CO or hydrocarbon. A second furnace may be required to heat up the catalyst precursor system to its dissociation temperature.

The PECVD emerged because certain processes could not tolerate the wafer temperatures of the thermal CVD. One noted problem was the charring of photo resists on patterned wafers. The PECVD provided an alternative in the microelectronic industry. The wafer temperatures required were lower from room temperature to 100°C. This became a key step in integrated circuit (IC) manufacturing. The precursor dissociation is enabled by high-energy electrons, allowing for low temperature operation. Researchers have attempted Direct Current (DC), Radio Frequency (RF), hot filament aided with DC, microwave, electron/cyclotron resonance, and inductively coupled plasma reactors.

Motorola[5] patented a solvent-based catalyst CVD method to grow CNTs in a reaction chamber. A nanotube structure is fabricated using a substrate, a mask region positioned on the substrate, patterning and etching through the mask region to form trenches, deposition of solvent-based nanoparticle catalyst onto the conductive material layer, removal of the mask region and subsequent layers grown using a lift-off process, and formation of nanotubes electrically connected to the conductive material layer. The nanotubes are formed from the catalyst using a reaction chamber with a hydrocarbon gas atmosphere. The substrate is made of silicon (Fig. 3.3). Other substrate materials such as glass, ceramic, metal, or semiconducting materials can be used. It can include electronic circuitry. A mask region is positioned on the surface of the substrate as shown in the sequence of fabrication steps in Fig. 3.3.

The mask region includes a bi-layer of a photo resist positioned on the surface. A bi-layer resist is used to facilitate the lift-off process. The mask region is patterned and etched to form at least one trench. An array of trenches is formed. The patterning is done through optical lithography, e-beam lithography, etc. A conductive material layer is deposited on the surface within a trench and a conductive material layer is deposited in another trench. These layers are gold (Au). Of course, aluminum, platinum, silver, or copper could be used as well. A solution containing a nanoparticle catalyst is deposited on the

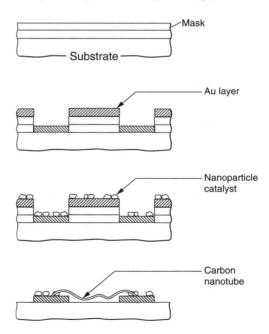

Figure 3.3
Sequence of steps in fabricating a CNT by CVD.

Mask

Substrate

Au layer

Nanoparticle catalyst

Carbon nanotube

conductive material layers. Nanoparticle catalysts include nanoparticles suspended within the solvent, which is compatible with the material included in the mask region. The nanoparticles are made of transition metals such as Fe, Ni, Co, etc. The catalysts may be deposited by several methods such as spraying, spinning, etc. A lift-off process is performed to remove the mask region from the substrate. The conductive layer with catalyst particles is removed during the lift-off. The nanotube structure is placed in a reaction chamber with a hydrocarbon gas atmosphere to form at least one nanotube. The reaction chamber is a CVD chamber or a molecular epitaxy chamber. Methane is used as the hydrocarbon gas atmosphere. Other gases such as ethylene, acetylene, or carbon monoxide can be used. As shown in Fig. 3.2, electrical connections can be affected. Thus, Motorola has patented an improved method of fabrication of a nanostructure. It uses a single-step patterning process. A bi-layer resist patterning process is used to facilitate the lift-off process. The process uses a solvent such as water. Contamination is minimized in the process.

The growth mechanism of CNT by CVD includes diffusion, adsorption, surface reaction, and desorption of the species. The essential steps in the mechanism are as follows:

1. Diffusion of the precursor
2. Adsorption of the species on the surface
3. Surface reactions
4. Desorption
5. Diffusion of the species

The energy needed for the surface reactions and desorption are provided by the bombardment of positive ions on the substrate. One or more of the previously mentioned five steps is the rate controlling steps. No careful experimentations have been reported on examining the rate-determining step as of yet. In the 1970s, studies on carbon filaments were reported. Hydrocarbons such as methane adsorb onto catalyst particles. Upon decomposition, carbon particles are released. These dissolve and diffuse into the particles. A supersaturated state is reached and then the carbon precipitates in a crystalline tubular form.

Two different models, i.e., the base growth model and the tip growth model, can describe the process from here on. In the base growth model, carbon precipitates from the top surface of the particle and the filament continues to grow with the particle anchored to the substrate. In the tip growth model, the particle attachment to the surface is weak. Carbon precipitation occurs at the bottom of the particle and the particle is lifted as it grows. The top end of the filament is decorated with catalyst particle. These mechanisms for carbon filament were based upon temperature-dependent growth rates, activation energy for various steps, and electron microscopy observations. The mechanisms for carbon filament growth by analogy may be applicable to the growth of CNT. The scale-up of the CNT processes to large scale would need a good understanding of growth mechanisms, the effect of process parameters on growth characteristics, gas phase, and surface kinetics. These would need modeling studies and experimentation

3.2.4   HIPCO Process

The HIPCO process was developed by Nobel laureate R.E. Smalley's team in 1998.[6] A gaseous catalyst precursor such as iron carbonyl is rapidly mixed with a flow of carbon monoxide (CO) gas in a chamber at high pressure and high temperature. The catalyst precursor decomposes and nanoscale metal particles form the decomposition products. These small particles serve as the catalyst. On the catalyst surface, CO molecules react to form carbon dioxide (CO_2) and carbon atoms, which bond together to form CNTs. This process is selective and 100 percent of the product is SWNTs. These CNTs can find applications in fuel cell electrodes, electronics, and biomedical applications.

SWNTs are produced selectively in the HIPCO process by gas-phase nucleation and growth from high-pressure CO. The product is free from contaminants and by-products. The transition metal catalyst is a gaseous phase catalyst. The reactant stream is maintained above the Boudouard reaction initiation temperature and the catalyst decomposition temperature. The CO gas stream and catalyst precursor are well mixed. Catalyst metal atom clusters are rapidly formed and are sufficient to promote the initiation and growth of SWNT to form a suspension of SWNT products in the gaseous stream. A high-pressure reaction vessel is used as shown in Fig. 3.4. The CNT diameter was found to be from 0.6 to 0.8 nm. CO at 30 atm pressure is

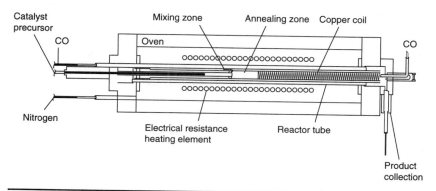

FIGURE 3.4   HIPCO apparatus.

preheated to approximately 1000°C. The catalyst precursor gas used is $Fe(CO)_5$. The growth and annealing zone is maintained at an elevated temperature of 1000°C. The tubes sometimes coalesce into ropes. The purity level of SWNTs is 99 percent. The diameter of SWNT can be controlled.

The HIPCO apparatus is shown in Fig. 3.4. The oven is a cylindrical aluminum pressure vessel containing an electrical resistance heating element surrounded by insulating material in the central portion. A reactor tube is suspended in axial orientation in the oven. The reactor tube includes both the mixing zone and the growth and annealing zone. The reactor tube is made of quartz and has a diameter of 7.5 cm and a length of 120 cm. An undiluted CO feed stream enters the oven near the exit and is passed counter-currently through a conduit at the periphery of the growth and annealing zone to supply CO to the mixing zone. The conduit is a copper coil of 0.25 in. in Outer Diameter, OD, spirally wound tubing. This configuration employs the heat in the quartz tube to preheat the CO gas stream fed to the mixing zone.

The source of carbon is CO. The catalytic mechanism that promotes the formation of SWNTs is not clear. Once critical aspect of the process is the formation of the active catalyst metal atom cluster of an appropriate size and initiation of SWNT growth. In order to form clusters of Fe atoms from dissociated precursor molecules such as $Fe(CO)_5$, the cluster must grow to the minimum nucleation size typically of 4 to 5 atoms. Aggregation is affected at this early stage by how tightly the initially formed Fe dimer is bound. The Fe dimer binding energy is relatively low, on the order of 1 electron volt. The formation of Fe atom aggregates of 4 to 5 atoms is a bit sluggish at the reaction temperatures of 800 to 1000°C. More rapid nucleation can be effected by including a nucleating agent in the gas feed stream. Such a nucleating agent can be a precursor moiety that under the reaction conditions stimulates clustering by decomposing more rapidly or binding to itself more tightly after dissociation. Atom clusters may be formed homogeneously or on seed clusters. The use of nucleating agents can increase the productivity of the process significantly up to 2 to 4 times. Increased partial pressures of CO facilitate faster cluster growth. SWNTs are formed by the Boudouard reaction. The mixing time is 1 msec.

The diameter of SWNT is proportional to the size of its active catalyst cluster at the time the tube starts to grow. The factors that control tube diameter of SWNT include the rate of aggregation of metal particles to form catalyst clusters and the rate at which nanotube growth begins upon a cluster of a given size. The relationship of these two rates can be controlled in three ways that can be used separately or together as desired. Larger ratios of the partial pressures of CO and gaseous catalyst result in smaller catalyst metal atom clusters, which provide smaller diameter tubes. An increase in the metal concentration will allow cluster formation to be more rapid, resulting in the formation of large tubes. The addition of a nucleating agent accelerates the aggregation rate of catalyst clusters and will result in an increase in the diameter of the tubes produced. Higher mixing zone temperatures result in smaller tubes.

3.2.5   Surface Mediated Growth of Vertically Aligned Tubes

Samsung[7] has patented a method for vertically aligning CNTs on a substrate. A CNT support layer is stacked on the substrate filled with pores. A self-assembled monolayer (SAM) is arranged on the surface of the substrate. The SAM includes an organic material containing phosphorous, such as 2-carboxyethyl phosphoric acid. The CNT support layer comprises a colloid monolayer with colloid particles such as silica or polystyrene (Fig. 3.5) and pores arranged between the colloid particles. The substrate includes a conductive material such as indium tin oxide (ITO).

A predetermined material layer is formed on the surface of a substrate so that one end of each of the CNTs can be attached well. The material layer is a SAM with a functional group having affinity toward the CNT. On the end of each of the CNTs are attached portions of the SAM exposed through the pores formed between the colloid particles. Since the lateral sides of the CNTs having large aspect ratios are supported by the colloid particles, the CNTs can be vertically aligned on the substrate having the SAM on it with the help of pores formed between the colloid particles. Silica particles of approximately 570 nm are dispersed in a propanol solution and then this solution is spin coated on the conductive substrate on which the SAM is formed so that the self-assembled colloid particles can be formed on the SAM. A second conductive substrate is arranged to be spaced at a predetermined distance from the first conductive substrate on which the

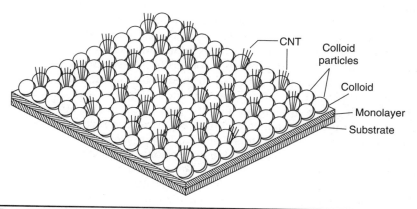

Figure 3.5   Vertically aligned CNT on a substrate.

colloid monolayer is formed. The second conductive material can be made transparent. The first and second conductive substrate materials are spaced 1 to 1.5 mm apart. The dispersion solution is injected between the two conductive substrates by capillary action.

When a predetermined anode voltage and cathode voltage are respectively supplied to the two conductive substrates, an electric field is generated between the two conductive substrates. One end of each of the CNTs contained in the dispersion solution is attached to the portions of the SAM exposed through the pores formed between the colloid particles by the electric field. Due to the functional group affinity to CNT, one end of each of the CNTs is stably attached to the SAM by chemical bonding. Further, the lateral sides of the CNTs having large aspect ratios are supported by the colloid particles so that the CNTs can be vertically aligned on the substrate having the SAM on it. Finally, when the dispersion solution and the second conductive substrate are removed, the CNTs remain vertically aligned through the pores on the substrate having the colloid monolayer on it as shown in Fig. 3.5. CNTs prepared in this manner are applied in variety of electronic devices such as field emission devices (FED).

3.3 Physical Properties of CNTs

CNTs were found to possess interesting physical properties. These are in the process of being discovered and characterized. The thermal conductivity observed in CNTs is of the order of 2000 W/m/K. The thermal conductivity varies with the diameter, chirality, and morphology of CNT. CNTs have unique electronic properties. They can be either metallic or semiconductor depending on their chirality (i.e., conformational variation). Many experiments and theoretical investigations have focused on electronic structures of CNTs in order to understand the origin of this remarkable phenomenon. In addition, large effort has been given to characterize their mechanical properties, such as Young's modulus, energetics, etc. To date, there is little progress made to characterize and to understand the thermal conduction in nanoscale materials. Some work has gone into understanding the lattice thermal transport properties for CNTs. Molecular dynamic simulation work has been reported to predict the higher values of thermal conductivity in CNTs. Thermal conductivity was found to change with change in temperature. It was difficult to measure the thermal conduction in nanostructures. Electrical conductance was measured in CNTs. Investigators report a quantum effect. This can be expected because of the comparable dimensions of the nanotube to the molecular diameter. Electrons can be expected to flow without hindrance through the CNT. Thus, where the obstacle effects are high, the free electron theory can be used to explain a possible increase in electrical conductivity in CNTs. The law of electrical resistivity for materials can be shown to vary inversely with the cross-sectional area of the resistor. When the area is in the order of nm^2, the resistivity can be predicted as having increased. However, the electron-confined transport inside the tubes would suggest a decrease in electrical resistivity. Careful experimentation is needed to measure the electrical properties of CNTs.

A parametric study on the dependence of electrical conductivity on the diameter, morphology, and chirality of CNTs needs be undertaken and general laws need to be determined that can be used in the nanorange. Large variations in the values reported for the elastic and tensile properties of CNTs can be found. The mechanisms of transport in CNTs have to be elucidated. Some investigators call the conduction phenomena in CNTs ballistic conduction.

CNTs exhibit dual property behavior. They can act as both conductors and semiconductors. A study was conducted on both the structure of CNTs and their corresponding electronic properties using two existing techniques. The two techniques would be used on each of the nanotubes studied, giving a complete picture of their unique structure and behavior as well as greater knowledge about how they "transition" from semiconducting to metallic in terms of their electronic properties. The work started at Columbia University, where the SWNTs were grown freely suspended over a slit etched into a silicon substrate. The researchers then identified usable individual nanotubes, labeled them, and studied them with a technique known as resonance Raleigh scattering.

This method allows detection of the optical spectrum of light scattered from the nanotubes and use of that scattered light to determine their electronic structure. The optical spectra alone, however, do not give sufficient information to absolutely assign electronic transitions to the physical structure of the nanotubes. A technique was needed that could provide independent structural verification. Investigators in the electron microscopy group at Brookhaven, NY, were interested in this problem and were able to provide electron diffraction as a solution.

Complementary data on physical structure of nanotubes were gathered using electron diffraction. Electron diffraction is an ideal tool for determining the exact structure of metallic and semiconducting nanotubes. They can use this tool to detect single-walled or double-walled nanotubes, and are not limited by the diameter range as compared with other methods. The theories of nanotube electronic transitions were tested and several assumptions made in previous models were confirmed. It has been verified how small changes in the pitch of the hexagons on the nanotube sidewall, determined by how the nanotube grows, lead to systematic deviations in the electronic behavior in both semiconducting and metallic structures. This predicted behavior, known as the "family pattern," had never before been directly tested, and the experimental results of Beetz and Sfeir[8] place it on a solid foundation that was previously lacking. Thus, they have measured the electronic structure of CNTs.

3.4   Applications

Potential applications for carbon nanostructures include microelectronics, scanning probes and sensors, FEDs such as video and computer displays, and nanoelectronics. Most open up promising near-term applications include electromagnetic shielding and electron field emission displays for computers and other high-tech devices and in applications requiring improved heat transfer and thermal insulation properties. Longer-range target applications include photovoltaics, super capacitors, batteries, fuel cells, computer memory, carbon electrodes, carbon foams, actuators, materials for hydrogen storage, adsorbents, and supports.

CNTs have such wide applicability due to their many unique mechanical, electrical, and chemical properties. These properties include electrical conductivity, mechanical strength, and thermal conductivity. For instance, CNTs may have mechanical strength of 10 to 100 times the strength of steel, but at a fraction of the weight. CNTs additionally demonstrate remarkably consistent electrical behavior. In fact, they exhibit an essentially metallic behavior and conduct electricity over well-separated electronic states while remaining coherent over the distances needed to interconnect various molecular computer components. Therefore, a wire produced from CNTs may potentially be used to connect molecular electronic components. Applications of CNTs include:

1. Electron field emitters in panel displays
2. Single-molecule transistors
3. Scanning probe microscope tips
4. Gas and electrochemical energy storage
5. Catalysts
6. Protein/DNA supports
7. Molecular filtration membranes
8. Energy absorbing materials
9. Hydrogen storage
10. Fuel cells
11. Super capacitors
12. Superconductors
13. Quantum conductors
14. Nanosensors
15. CNT-based composites

It is realized that the atomic arrangement in a CNT and its electrical properties may vary drastically along the length of the nanotube.[9] Such a variation in electrical properties may adversely affect the efficiency of electron transport between nanodevices interconnected by the CNT.

3.5   Morphology of CNTs

Carbon nanostructures can be produced with different morphologies. Some examples of different morphologies of CNTs are:

SWNTs
DWNTs
MWNTs
Nanoribbon
Nanosheet
Nanopeapods
Linear and branched CNTs
Conically overlapping "bamboo-like" tubule
Branched Y-shaped tubule
Nanorope
Nanowire
Nanofiber

Inorganic nanopeapods were grown at the Max Planc Institute of Microstructure Physics[10] in Halle, Germany. Facile control of both the size and separation of platinum nanoparticles within a $CoAl_2O_4$ nanoshell was possible. Nanowires with alternating layers of cobalt and platinum are electrodeposited within a nanoporous anodic aluminum oxide membrane. Annealing the membrane at 700°C allows a reaction between cobalt and alumina, forming continuous $CoAl_2O_4$ pods. The heating also causes the platinum to agglomerate into spherical "peas" within these shells. The lengths of the cobalt and platinum segments determine the diameter and distance between each of the platinum peas. Long cobalt segments lead to large spaces between the particles, whereas short cobalt segments form more tightly packed peas.

Strong van der Waals attraction forces allow for spontaneous roping of nanostructures leading to the formation of extended carbon structures. A SWNT has only one atomic layer of carbon atoms. A DWNT has two atomic layers of carbon atoms and an MWNT comprises up to one thousand cylindrical graphene layers concentrically folded about the needle axis. MWNTs have excellent strength, small diameter (less than 200 nm), and near metallic electrical conductivity as well as other interesting properties. They can be used as

additives to enhance structural properties of carbon-carbon composites, carbon-epoxy composite, metal-carbon composites, and carbon-
concrete composites. The properties of these materials depend on
the topology, morphology, and quality of the nanotubes. In a flat
panel display application, the alignment of the tubes is important.
There is sufficient interest in developing processes that can generate
nanotubes with predetermined morphology in an economic and
environmentally safe manner. There is an identified need for a nanostructure having a high surface area layer containing uniform pores
with a high effective surface area. This would increase the number of
potential chemical reactions or catalyst sites on the nanostructure.
These sites may also be functionalized to enhance chemical activity.
Nanofibers are needed with large surface area filled with additional
pores and more functionality with increased number of reaction/
catalyst sites.

Matyjaszewski et al.[11] patented a novel and flexible method for
the preparation of CNTs with predetermined morphology. Phase separated copolymers/stabilized blends of polymers can be pyrolyzed
to form the carbon tubular morphology. These materials are referred
to as precursor materials. One of the comonomers that form the
copolymers can be acrylonitrile (AN), for example. Another material
added along with the precursor material is called the sacrificial material. The sacrificial material is used to control the morphology, self-
assembly, and distribution of the precursor phase. The primary source
of carbon in the product is the precursor. The polymer blocks in the
copolymers are immiscible at the micro scale. Free energy and entropic considerations can be used to derive the conditions for phase
separation. Lower critical solution temperature (LCST) and upper
critical solution temperature (UCST) are also important considerations in phase separation of polymers. However, they are covalently
attached, thus preventing separation at the macroscale. Phase separation is limited to the nanoscale. The nanoscale dimensions typical
of these structures range from 5 to 100 nm. The precursor phase pyro-
lyzes to form carbon nanostructures. The sacrificial phase is removed
after pyrolysis.

When the phase-separated copolymer undergoes pyrolysis, it
forms two different carbon-based structures such as a pure carbon
phase and a doped carbon phase. The topology of the product
depends on the morphology of the precursor. Due to the phase separation of the copolymer on the nanoscale, the phase-separated copolymers self-assemble on the molecular level into the phase-separated
morphologies. ABC block copolymer may self-assemble into over
20 different complex phase-separated morphologies (Fig. 3.6 and
Fig. 3.7). Typical morphologies are spherical, cylindrical, and lamellar.
Phase-separated domains may also include gyroid morphologies
with two interpenetrating continuous phases. The morphologies
are dependent on many factors such as volume ratio of segments,

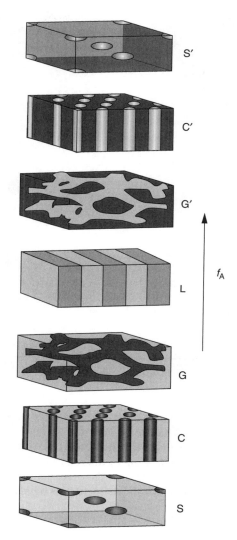

FIGURE 3.6
Different morphologies of precursor material—biphasic.

chemical composition of segments, connectivity, Flory-Higgins inter-action parameters between segments, and processing conditions.

The nanoscale morphologies formed by spontaneous or induced self-assembly of segmented copolymers exist below a certain critical temperature. The morphologies are reversibly formed. When morphology with cylindrical domain is selected as precursor, the product is made of CNTs. Oriented nanostructures can be formed. The thickness of the film used to affect the process affects the length of the cylinder. Spinning or extrusion of the polymer at the surface prior to carbonization leads to nanowires used in nanoelectronics after

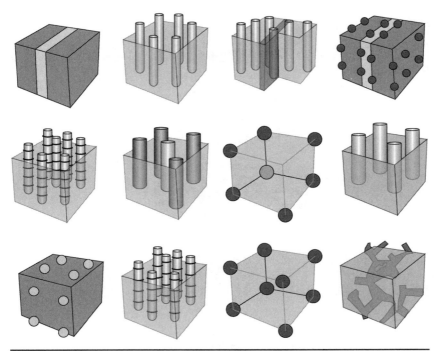

Figure 3.7 Different morphologies of precursor material—triphasic.

pyrolysis. Nanotubes, nanowires, and nanofibers can be formed in this manner (Fig. 3.8).

Poly-acrylonitrile (PAN) has been used in the industry to form carbon fibers. Cross-linking of micro-scale phase-separated domains in the polymer blend that forms the precursor can lead to the formation of nanoclusters in the product. Ti clusters have a lot of applications in semiconductors.

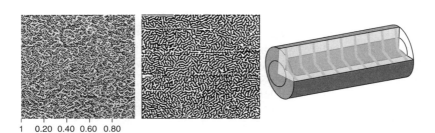

1   0.20 0.40 0.60 0.80

Figure 3.8 Microscopic images of brush copolymers with grafts.

The precursor polymer blend can be made out of:

1. Linear block copolymers
2. AB block copolymers
3. ABA block copolymers
4. ABC block copolymers
5. Multiblock copolymers
6. Symmetrical and asymmetrical star block copolymers
7. Blends of polymers
8. Graft copolymers
9. Multibranched copolymers
10. Hyperbranced or dendritic block copolymers
11. Novel brush copolymers

The blocks of each of the polymers may be comprised of:

1. Homopolymer block
2. Statistical polymer block
3. Random copolymer block
4. Tapered copolymer block
5. Gradient copolymer block
6. Other forms of block copolymers

Any copolymer topology that allows immiscible segments to be present in the system to phase separate inherently or upon annealing either alone or in the presence of a solvent is a suitable precursor material.

3.6   Summary

CNTs are rolled graphene sheets of atoms about their needle axis. They are 0.7 to 100 nm in diameter and a few microns in length. Carbon hexagons are arranged in a concentric manner with both ends of the tube capped by a pentagon containing a Buckminster fullerene-type structure. They possess excellent electrical, thermal, and toughness properties. Young's modulus of CNT has been estimated at 1 TPa with a yield strength of 120 GPa. Iijima verified fullerenes in 1991 and observed MWNTs formed from carbon arc discharge.

Five methods of synthesis of CNTs are discussed. These are: (1) arc discharge, (2) laser ablation, (3) CVD, (4) HIPCO process, and (5) surface mediated growth of vertically aligned tubes. NEC developed the arc discharge process in 1992. Two graphite rods were connected to a power supply spaced a few millimeters apart. At 100 amps, carbon vaporizes and forms hot plasma. Typical yields are 30 to 90 percent. The SWNTs and MWNTs are short tubes with diameters of 0.6 to 1.4 nm. They can be synthesized in open air. The product needs purification. The CVD process was invented by Nagano, Japan. The substrate is placed in an oven, heated to 600°C and a carbon-bearing gas such as methane is slowly added. As the gas decomposes, it frees up the carbon atoms, which recombine as a nanotube. Yield range is 20 to 100 percent. Long tubes with diameter ranging from 0.6 to 4 nm were formed. It can be easily scaled up to industrial production. The SWNT diameter is controllable. The tubes are usually multiwalled and riddled with defects. Smalley developed the laser vaporization process in 1996. Graphite is blasted with intense laser pulses to generate carbon gas. Prodigious amounts of SWNTs are formed. Yield of up to 70 percent is found. Long bundles of tubes 5 to 20 μm with diameters in the range of 1 to 2 nm are formed. The product formation is primarily SWNTs. Good diameter control is possible and few defects are found in the product. Reaction product is pure. The process is expensive.

Smalley also developed the HIPCO process in 1998. A gaseous catalyst precursor is rapidly mixed with CO in a chamber at high pressure and temperature. Catalyst precursor decomposes and nanoscale metal particles form the decomposition product. CO reacts on the catalyst surface and forms solid carbon and gaseous CO_2. The carbon atoms roll up into CNTs. One hundred percent of the product is SWNT and the process is highly selective. Samsung patented a method for vertically aligning CNTs on a substrate. A CNT support layer is stacked on the substrate filled with pores. SAM is arranged on the surface of the substrate. On the end of each CNT are attached portions of the SAM exposed through the pores formed between the colloid particles present in the support layer. CNTS can be vertically aligned on the substrate having the SAM on it with the help of pores formed between the colloid particles.

CNTs possess interesting physical properties. Thermal conductivity of CNTs is in excess of 2000 w/m/K. They have unique electronic properties. Applications include electromagnetic shielding, electron field emission displays for computers and other high-tech devices, photovoltaics, super capacitors, batteries, fuel cells, computer memory, carbon electrodes, carbon foams, actuators, material for hydrogen storage, and adsorbents.

CNTs can be produced with different morphologies. Examples of different morphologies include SWNT, DWNT, MWNT, nanoribbon, nanosheet, nanopeapod, linear and branched CNTs, conically overlapping bamboo-like tubule, branched Y-shaped tubule, nanorope, nanowires, and nanofilm. Processes are developed to prepare CNTs with the desired morphology. Phase-separated copolymers/stabilized blends of polymers can be pyrolyzed along with sacrificial material to form the desired morphology. The sacrificial material is changed to control the morphology of the product. Self-assembly of block copolymers can lead to 20 different complex phase-separated morphologies. Often times as is the precursor, so is the product. Therefore, even more variety of CNT morphologies can be synthesized.

Review Questions

1. What is the relation between CNTs and fullerenes?

2. How are nanowires drawn from CNTs?

3. Is there a relation between chirality of graphitic rings and semiconductor properties of CNTs?

4. What is meant by straight form of CNTs?

5. How are force field calculations used to predict CNT behavior?

6. What is the role of AFM during the preparation of CNTs?

7. What is the McGroddy Prize and ACS medal?

8. What is meant by soot harvesting?

9. What is the degree of control available in the electric arc process for control of diameter of CNTs?

10. What is the degree of control available in the laser ablation process for control of diameter of CNTs?

11. What is the degree of control available in the CVD process for control of diameter of CNTs?

12. What is the degree of control available in the HIPCO process for control of diameter of CNTs?

13. What is the degree of control available in the surface mediated process for vertical alignment of CNTs?

14. What is the typical length of CNT formed during the electric arc process for synthesis of CNTs?

15. What is the typical length of CNT formed during the laser ablation process for synthesis of CNTs?

16. What is the typical length of CNT formed during the CVD process for synthesis of CNTs?

17. What is the typical length of CNT formed during the HIPCO process for synthesis of CNTs?

18. What is the typical length of CNT formed during the surface mediated vertically aligned nanotubes process for synthesis of CNTs?

19. What are the typical temperature and pressure used during the synthesis of CNTs using the electric arc process?

20. What are the typical temperature and pressure used during the synthesis of CNTs using the laser ablation process?

21. What are the typical temperature and pressure used during the synthesis of CNTs using the CVD process?

22. What are the typical temperature and pressure used during the synthesis of CNTs using the HIPCO process?

23. What are the typical temperature and pressure used during the synthesis of CNTs using the surface mediated vertically aligned nanotubes?

24. What is the difference between SWNT, DWNT, and MWNT?

25. What is the role of catalyst in the laser ablation process?

26. What is the time duration of laser pulse during laser ablation process to prepare CNTs?

27. What does pelletization do in the laser ablation process?

28. Rank the CVD, laser ablation, and electric arc processes to form CNTs.

29. What does the quartz tube do in the CVD process to form CNTs?

30. Mention the typical growth rates of CNTs in the CVD process.

31. Discuss the sequence of steps in the CVD process to prepare CNTs.

32. Sketch the CVD method showing clearly the mask, substrate, deposition of solvent based catalyst, removal of mask region, etc.

33. What is unique about the "lift-off" process?

34. What is the role of trench formation and etching in the CVD process?

35. What is the role of diffusion in the growth mechanism of CNT by CVD?

36. What is the energy needed for surface reactions and desorption for CNT synthesis?

37. Comment on the rate determining step among diffusion, adsorption of species on surface, surface reactions, desorption, and diffusion of species during CNT synthesis by CVD.

38. What are the typical operating conditions during the HIPCO process developed by Smalley in 1998?

39. What is the purpose of the annealing zone in the HIPCO process?

40. Discuss gas-phase nucleation and growth during the HIPCO process.

41. Discuss the formation of SWNTs by Boudouard reaction.

42. Distinguish the substrate layer from the material layer during the surface mediated vertical alignment of CNTs.

43. Compare the thermal conductivity of CNTs with that of steel.

44. Compare the Young's modulus of elasticity of CNTs with that of steel.

45. Compare the yield strength of CNTs with that of steel.

46. What are the 12 different CNT morphologies identified?

47. How many different morphologies does a "phase-separated copolymer blend" exhibit?

48. How is the predetermined morphology of a CNT structure obtained by the selection of raw materials?

49. What is the expected application of CNT in a solar power plant?

50. What is the expected application of CNT in biomedical engineering?

References

1. K. R. Sharma, Control of Diameter during CNT Synthesis by the Three Methods, 99th AIChE Annual Meeting, Salt Lake City, UT, November, 2007.
2. H. Huang, H. Kajiura, M. Miyakoshi, A. Yamada, and M. Shiraishi, Arc Electrodes for Synthesis of Carbon Nanostructures, US Patent 6,794,598, 2003, Sony Corp., Tokyo, Japan.
3. M. Yudasaka and S. Iijima, Process for Producing Single Wall Carbon Nanotubes Uniform in Diameter and Laser Ablation Apparatus used Therein, US Patent 6,331,690, NEC Corp., Tokyo, Japan.
4. M. Meyyappan, Carbon nanotube growth by chemical vapor deposition, in *Encyclopedia of Nanoscience and Nanotechnology*, H.S. Nalwa, Ed., 2004, pp. 581–589, American Scientific Publishers, Valenica, CA.
5. R. Y. Zhang, R. K. Tsui, J. Tresek, and A. M. Rawlett, Method for Selective Chemical Vapor Deposition of Nanotubes, US Patent 6,689,674, 2004, Motorola, Schaumburg, IL.
6. R. E. Smalley, K. A. Smith, D. T. Colbert, P. Nikolaev, M. J. Bronikowski, R. K. Bradley, and F. Rohmund, Single-Wall Carbon Nanotubes from High Pressure CO, US Patent 7,204,970, 2007, Rice University, Houston, TX.
7. Y. W. Jin, J. M. Kim, H. T. Jung, T. W. Jeong, and Y. K. Ko, Carbon Nanotube Structure and Method of Vertically Aligning Nanotubes, US Patent 7,371,696, 2008, Samsung SDI Co. Ltd., Suwon-si, Gyeonggi-do, KR.
8. M. Y. Sfeir, T. Beetz, F. Wang, L. Hwang, X. M. H. Huang, M. Huang, J. Hone, et al., "Optical spectroscopy of individual single-walled carbon nanotubes of defined chiral Structure", Science, 312, 5773, 554–556, 2006.
9. P. G. Collins, A. Zettl, H. Bando, A. Thess and R. E. Smalley, "Nanotube nanodevice", Science, 278, 5335, 100–102, 1997.
10. Science and technology concentrates, *Chemical & Engineering News*, 86, 31, 36–37, 2008.
11. K. Matyjaszewski, T. Kowalewski, D. N. Lambeth, J. T. Spanswick, and N. V. Tsarevsky, Process for the Preparation of Nanostructured Materials, US Patent 7,056,455, 2006, Carnegie Mellon University, Pittsburgh, PA.

CHAPTER **4**

Nanostructuring Methods

Learning Objectives

- "Top-down" and "bottom-up" strategies
- Nanowires by vacuum synthesis, nanoparticles by gas evaporation, nanoprisms using light, nanorods by condensation
- Subtractive and additive fabrication; lift-off processes
- Quantum dots (QD), self-assembled monolayer (SAM)
- Sol-gel processing, dry etching, reactive ion etching
- Dip-pen lithography, nanoimprint lithography, electron beam lithography, atomic lithography, and direct-write lithography
- Galvanic fabrication, hot embossing, nanomechanical method
- Layered morphology in polymer thin films
- Cryogenic milling, electrodeposition, plasma compaction
- Nanofluids, pulsed laser deposition
- Self-assembly of copolymers and predetermined morphology formation

Nanostructures can be nanowires, nanorods, branched nanowires, nanotetrapods, nanocrystals, QDs, nanosheets, nanocylinders, nanocubes, nanograins, nanofilaments, nanolamella, nanopores, nanotrenches, nanotunnels, nanovoids, and nanoparticles. A nanostructure can be defined as a structure with at least one or two dimensions in the 1 to 100 nm range. It can be prepared as dispersion in a second phase material. Nanoparticles can be synthesized in a variety of methods. Many processes are available for the manufacture of small metal particles. These processes cover a wide range of technologies and exhibit a wide range of efficiencies. Some processes produce dry particles, while other produce particles in liquid dispersions. Nanostructures

"Top-Down" Semiconductor Nanoscale Technology	"Bottom-Up" Molecular Nanotechnology
• From big (bulk wafer) to small (chip)	• From small (self-assembled structure) to big (chip)
• "Pattern" and "etch"	• "Synthesis" and "self-assembly"
• Expensive fabrication	• Cheap fabrication
• Less scalable material/device	• More scalable material/device
• Inflexible in material selection	• Flexible in material selection
• Matured design tool/infrastructure	• Open design tool/infrastructure
• "Hit the red-brick wall very soon"	• Leads to "molecular level engineering"

TABLE 4.1 Comparison between Two Nanotechnologies

can either be generated by building from atoms (called the bottom-up strategy) or by diminishing the size from microparticles to nanoparticles (called the top-down strategy). In Table 4.1, the two methods are compared.

4.1 Vacuum Synthesis

Quasi one-dimensional solid-state nanostructures such as nanowires can be synthesized by sputtering a uniform flow of nitrogen molecular ions under ultrahigh vacuum. Periodic wave-like patterns are formed[1] with troughs level with the silicon-insulator border of the silicon-on-insulator (SOI) material (Fig. 4.1). The ion energy, the ion incidence

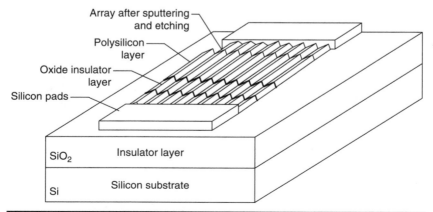

FIGURE 4.1 FET device formed by sputtering and etching under high vacuum.

angle to the surface of material, the temperature of the silicon layer, the depth of the wave-like pattern, the height of the wave-like pattern, and the ion penetration range in the silicon are determined based on a selected wavelength in the range of 9 to 120 nm. These nanostructures are employed in optoelectronic and nanoelectronic devices such as FET.

SOI can be made using SIMOX technology. The silicon nitride layer is deposited on top of a thin silicon oxide layer. Lithography and plasmochemical etching are used to form the mask window. The thin oxide layer within the mask is removed by wet chemical etching. A pendant edge is formed around the periphery of the mask window. The mask window is oriented toward the ion beam as desired for efficient sputtering. Sputtering is carried out under ultrahigh vacuum based on E, T, and θ values. Fabrication at 850 K results in wave patterns different from that obtained at room temperature. The wavelength is reduced by a factor of 3.3. The thickness of the layer and slope of the sides of the waves remain the same at both temperatures. The sputtering process is followed by an annealing process at a temperature of approximately 1000 to 1200°C in an inert environment for a period of 1 h followed by high temperature oxidation. Crystalline silicon is made to form into isolated quantum wires. Fabrication of an FET device by such a method is shown in Chap. 2, Fig. 2.1. The dimensions in such devices are smaller than those made before.

4.2   Gas Evaporation Technique

The gas evaporation technique is used to produce ultrafine metal powders, especially magnetic metal or metal oxide powders. These materials are also referred to as magnetic pigments. The process is a dry process and does not involve any contact with liquids. Sometimes the metal is evaporated onto a thin film of hydrocarbon oil and is called the VEROS technique. In another process, surface-active agents stabilize a dispersion of a ferromagnetic metal such as Fe, Co, or Ni vaporized directly into a hydrocarbon oil to give a ferrofluid using a metal atom technique. The metal atom technique requires high vacuum pressure of less than 1000 torr such that the metal atoms impinge onto the surface of a dispersing medium. Nucleation and particle growth occur in the dispersing medium. The particle size depends on the dispersing medium and cannot be easily controlled. Another process for making magnetic fluids involves vaporization of a ferromagnetic metal, adiabatic expansion of the metal vapor and an inert gas through a cooling nozzle to condense the metal and form small metal particles, and impingement of the particles at high velocity onto the surface of a base liquid. Colloidal metal dispersions in nonaqueous media can be prepared by nanomechanical dispersion of fine metal particles. Nanoparticles from organic compounds with improved stability toward flocculation in the size range of 500 A to 4 μm can be prepared by the gas evaporation technique.

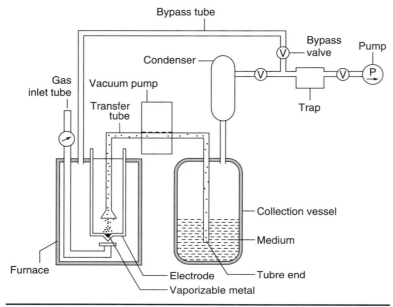

FIGURE 4.2 Apparatus to produce metal nanoparticles by gas evaporation.

In Fig. 4.2[2] an apparatus patented by Cima Nanotech, Woodbury, Minnesota, to prepare metallic nanopowders using the gas evaporation technique is shown. Small particles of metals are prepared by an evaporative method with a unique collection method that increases the production efficiency of the process by dramatic degrees. The process comprises evaporating a metal and then providing a mechanical pump that either draws the gas phase metal into a liquid condensation-collection zone or combines a liquid condensation-collection zone within the mechanical pump. The nonmetal gaseous material remaining after condensation removal of the metal material is withdrawn from the material stream, while the liquid condensing phase with the condensed metal particles is separated, the liquid condensing phase carrier removed, and the particles collected. As compared to known prior art methods, the use of the intermediate positioned mechanical pump or contemporaneous mechanical pump and condensation-collection zone increases the overall collection/manufacturing efficiency of the process by at least 25 percent.

A source of nanoparticles is provided. The source may be a primary source where particles are being manufactured (e.g., sputtering, spray drying, aerial condensation, aerial polymerization, and the like). The source of nanoparticles may also be a secondary source of particles, where the particles have been previously manufactured and are being separately treated (e.g., coating, surface oxidation, surface etching, and the like). These nanoparticles are provided in a gaseous medium that is of a sufficient gas density to be able to support the particles in flow. That is, there must be sufficient gas that when the gas is moved, the particles will be carried. With nanoparticles (particles having number average diameters of 1 to 100 nm, preferably 1 to 80 nm, or 1 to 70 nm, and as low as 1 to 50 nm), only a small gas pressure is needed, such as at least 0.25 torr, although higher pressures greater than 0.25, 0.4, 0.6, 0.75, and 0.9 torr are preferred.

The gas-carrying medium may be reactive with the particles or may have some residual reactive materials in the gas. It is preferred that the gas is relatively inert to the apparatus environment. Gases such as nitrogen, carbon dioxide, air, and the like are preferred. The propulsion system for the gas-carrying medium and the nanoparticles is a dry mechanical pumping system for gases. A dry pumping system is used to prevent contamination of the particles by lubricants. These dry pumping systems for gases are well known in the semiconductor industry for conveying air, particulate, and vapors without collection occurring in the pump. They are pumping systems that utilize oil-less seals to maintain vacuum conditions at the pump inlet.

A particle collection system with increased collection efficiency for the collection of nanoparticles comprises a source of particles, a dry pumping system, and a particle collection surface. The position of a dry pumping system in advance of the particle collection surface maintains a particle moving effort, without wetting particles and causing them to agglomerate, and increases collection efficiency.

The placement of the collection units between the nanoparticle source and vacuum pumps causes severe problems in maintaining the system vacuum and related high evaporation rates. Wet collection systems are difficult to operate in a vacuum environment; however, the operation of wet collection systems provides slurries in a number of different solvents, which can be posttreated by in situ polymerization techniques to coat the nanoparticles. The particles in the resulting slurries can be coated with fluoropolymers, such as Teflon and polyvinylidene difluoride (PVdF) by in situ polymerization methods. This differs from earlier work by the use of high-pressure reactor technology to provide a Teflon or PVdF coating onto the particle. This is the first known application of these polymers in an in situ polymer coating process.

The collecting medium for the nanoparticles may comprise electrostatic surface collectors, electrostatic filter collectors, porous surfaces (e.g., fused particle surfaces), centrifugal collectors, wet scrubbers, liquid media collectors, and physical filter collectors. The liquid media collectors with subsequent separation of the liquid and the particulates are more amenable in the practice of the present invention. Also known as wet scrubbers, these liquid collection media are more amenable to this arrangement due to process and safety factors, allowing more volatile solvents to be utilized away from the formation chamber for the nanoparticles. Wet scrubbers also provide slurries suitable for posttreatment and polymer coating by in situ polymerization, particularly in the case of fluoropolymer coatings. Examples of this are Teflon, PVdF, and their respective copolymers.

The use of the present arrangement of nanoparticle source, dry pump, and collector has been found to increase particle collection efficiency by as much as 100 percent in comparison to the conventional source, a filter pump system, even where the same nanoparticle source is present, the same filter and the same pump is used in the different order. The utilization of this arrangement of the pumping scheme may also benefit the collection of the nanoparticles. By injecting low volatility solvents into the inlet of the pump with the nanoparticle loaded gas stream, the dry pump may also be utilized as a wet scrubber with better than 90 percent collection efficiency.

4.3   Triangular Nanoprisms by Exposure to Wavelength of Light

Nanoclusters are an important class of materials that are having a major impact on a diverse range of applications, including chemical and biodetection, catalysis, optics, and data storage. The use of such particles dates back to the Middle Ages, and the scientific study of them has spanned over a century. These nanostructures are typically made from molecular precursors, and there are now a wide variety of compositions, sizes, and even shapes available. Because of their unusual and potentially useful optical properties, nanoprism structures in particular have been a recent synthetic target of many research groups. A high-yield photosynthetic method for the preparation of triangular nanoprisms from silver nanospheres was recently reported.[3] For many nanoparticle syntheses, an Ostwald ripening mechanism, where large clusters grow at the expense of smaller ones, is used to describe and model the growth processes. This type of ripening typically results in unimodal particle growth. Thus, a method of controlling the growth and ultimate dimensions of such structures is desired. Such a method will necessarily fall outside of the known Ostwald ripening mechanisms.

A method of forming nanoprisms by exposing silver particles to a wavelength of light between approximately 400 and 700 nm for a period of less than approximately 60 h was proposed.[3] The nanoprisms formed have a bimodal size distribution. The silver particles are present in a colloid containing a reducing agent, a stabilizing agent, and a surfactant. If the colloid contains a stabilizing agent and a surfactant, the ratio of the stabilizing agent to the surfactant is preferably about 0.3:1. The nanoparticle starting materials have a diameter between 0.2 nm and approximately 15 nm. The nanoprisms formed are single crystalline and have a {111} crystal face on a base plane of the nanoprism and a {110} crystal face on a side plane of the nanoprism

S. No	Primary Wavelength of Light (nm)	Secondary Wavelength of Light (nm)	Nanoprism Edge Length (nm)
1.	340	450–700	31–45
2.	470–510		53–57
3.	500–540		53–71
4.	530–570		64–80
5.	580–620		84–106
6.	650–690		106–134

TABLE **4.2**   Nanoprism Edge Length Variation with Primary and Secondary Wavelength of Light

and display plasmon bands having λ_{max} at 640 and 1065 nm, and 340 and 470 nm, respectively.

Another method of forming a nanoprism by exposing silver nanoparticles to a primary and secondary wavelength of light such that one of the primary and secondary wavelengths of light excites quadrupole plasmon resonance in the silver particles was also presented. One of the primary and secondary wavelengths of light coincides with the out-of-plane quadrupole resonances of the silver nanoprisms. The secondary wavelength of light is approximately 340 nm and the primary wavelength of light is in the range of approximately 450 and 700 nm. By adjusting the primary wavelength of light, the edge length of the nanoprisms produced can be controlled as shown in Table 4.2.

4.4   Condensed Phase Synthesis

Nanorods were synthesized in condensed phase by a process described by UT-Battelle.[4] Conversion and growth were effected in the solid phase. Spectroscopic data were acquired to control the process. Carbon/silicon is deposited by condensation of gaseous phase carbon/silicon. The formation rates of nanostructure by this method allows for reasonable production rates.

CNTs can be grown from heated mixtures of carbon and catalyst powders. Pulsed laser vaporization was used at various pressures. Different sizes of carbon and metal catalyst particles are codeposited in this method. In situ diagnostics are accomplished by TEM, AFM, and field emission scanning electron microscopy (FESEM). Powder mixtures were heated in vacuum with background gases using resistive heating elements and auxiliary laser irradiation or plasma spray excitation. TEM investigation facilitated nanotube growth. A major technical milestone is the demonstration of growth of nanotubes from mixed powders.

A plasma torch is used to melt micron-sized particles. The plasma torch is powder fed. The melted particles are directed at high velocity toward a substrate. Molten particles are accelerated toward the substrate where they impact, splat, and cool rapidly. This constitutes a normal plasma spray. Simultaneous deposition of particles, heating of the substrate, and resolidification of the deposit is achieved by the torch. The cluster size of nanoparticles is allowed to increase in order to achieve high deposition rates desirable for rapid conversion to nanotubes and high volume CNT growth.

The carbon feed material and catalyst material is vaporized to form atomic carbon. Nanoscale particles are allowed to form after an appropriate time interval and later compared to the carbon particles, which agglomerate and cluster (Fig. 4.3). The catalyst may be metallic particles or aggregates of them. The CNT is formed after heat treatment, sintering, and annealing. The particle-based supply is at a high rate sufficient to maintain the growth of the CNTs. Deposition is performed by a method to allow and achieve directed growth. Directed growth is performed by a method that allows for the specific shape and formation of a specific component geometry and structure.

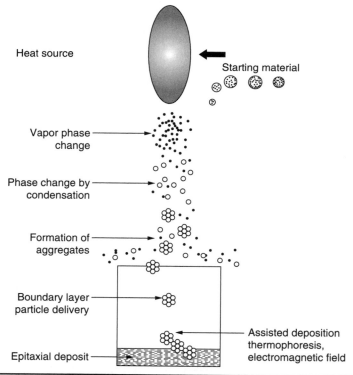

Heat source

Starting material

Vapor phase change

Phase change by condensation

Formation of aggregates

Boundary layer particle delivery

Assisted deposition thermophoresis, electromagnetic field

Epitaxial deposit

FIGURE 4.3   Condensed phase synthesis of nanostructure.

Diffusion limited transport is overcome by particle inertia and particle-assisted delivery to supply material to the substrate at high rates. Rapid epitaxy is achieved by auxiliary heating.

Thus, the process comprises three essential steps: (1) growth of CNT from mixed particles, (2) deposition of nanocarbon at particle delivery rates suitable for a modified spray technique, and (3) solid-state conversion into CNTs. Transient heating is all that is required to form CNTs by heat treatment.

4.5   Subtractive and Additive Fabrication

Nanostructures can be created by subtractive and additive fabrication. During subtractive fabrication, the requirements for the process removal of the functional material depend on the thickness of the material. Nanometer structures with low aspect ratios are possible. Wet chemical and plasma chemical methods of subtractive patterning and isotropic gas phase process can be used for this purpose. A resist layer that is structured by a lithographic process covers an ultrathin functional layer. Lithographic masks are applied in an analogous fashion from microtechnological lithography. Pattern transfer is affected into the ultrathin layer prior to the removal of the mask. This subtractive patterning of the ultrathin layer is comparable to microtechnological etching. This is because the process is based upon a series of chemical reactions that effect a removal of the layer in the apertures of the mask and transfer of the material into the gas phase. Process times are in the second to minute range. Hence, the removal of ultrathin layers requires only a low rate of removal. Sometimes the removal of the ultrathin layers is effected in a single step. This is the case for cleavage of a small group of atoms or the local removal of molecules from a molecular monolayer (Fig. 4.4). Nanolocal surface manipulation is effected in the process. This is a surface chemistry guided process. This is not limited to the removal of molecules or atoms. Binding of new atoms or a group of atoms to the surface is also possible. As shown in Fig. 4.5 such a synthetic process is an additive pattern transfer. New material is added during the process. This can be extended to several layers of atoms or molecules. The steepness of the mask edges has to be sufficient in the case of thick masks to facilitate lateral structures in the nanoscale range. Galvanic fabrication is an example. For a monolayer, the distinction between additive and subtractive structures is minimal. Only when additional layers of molecules and atoms are considered can the additive fabrication technique be seen to add material through the window of the mask. In a similar fashion, the material is removed from the window of the masks in the subtractive fabrication procedure.

In the fabrication of an IC circuit and nanowires, lift-off processes can be used to generate the nanostructure. In Fig. 4.6[5] a flow chart for preparation of the iridium oxide (IrOx) nanostructure is shown. It has

Figure 4.4 Local oxidation of alkylated surfaces in the windows of the mask during fabrication of lateral microstructured molecular monolayers by lithography.

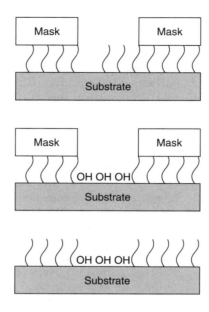

Figure 4.5 Local alkylation of an OH group rich surface in the windows of a mask during fabrication of lateral microstructured molecular monolayers by lithography.

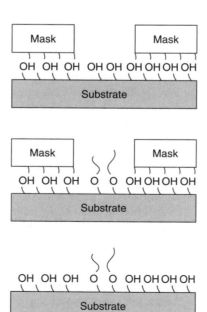

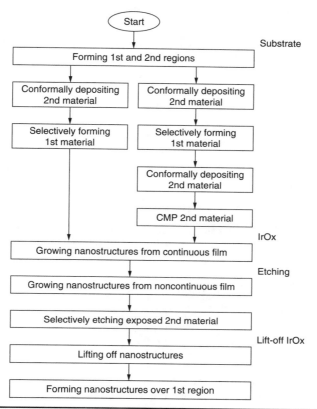

FIGURE 4.6   Flowchart of patterning IrOx nanostructure.

been shown in the literature that nanotips and nanorods can be efficiently formed using conventional complementary metal-oxide-semiconductor (CMOS) processes. As a natural next step, a method was developed for forming practical IrOx nanotip structures. A process for patterning IrO_2 nanorods, so that they can be seamlessly integrated into CMOS, IC, and liquid crystal display (LCD) devices, was developed. A method was described for patterning IrOx nano-structures. The method comprises the following:

1. Forming a substrate first region adjacent to a second region

2. Growing IrOx nanostructures from a continuous IrOx film overlying the first region

3. Simultaneously growing IrOx nanostructures from a non-continuous IrOx film overlying the second region

4. Selectively etching areas of the second region exposed by the noncontinuous IrOx film, lifting off the IrOx nanostructures overlying the second region

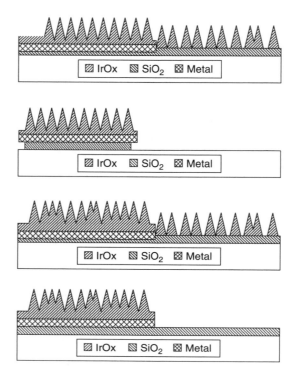

5. In response to lifting off of the IrOx nanostructures overlying
    the second region, forming a substrate with nanostructures
    overlying the first region

Typically, the first region is formed from a first material and the
second region from a second material, different from the first material
(Fig. 4.7). For example, the first material can be a refractory metal or
refractory metal oxide. The second material can be SiOx.

The step of selectively etching areas of the second region exposed
by the noncontinuous IrOx film includes exposing the substrate to an
etchant that is more reactive with the second material than the IrOx.
For example, if the first material is a refractory metal and the second
material is SiO2, then hydrogen fluoride (HF) or buffered oxide etches
(BOE) are suitable etchants.

The steps of forming a substrate first region adjacent to a second
region include the following: conformally depositing the second material overlying the first and second regions, and selectively forming the
first material overlying the second material in the first region. In a
second aspect, the step of forming a substrate first region adjacent to a
second region includes the following: conformally depositing the second
material overlying the first and second regions, selectively forming the
first material with a surface overlying the second material in the first
region, conformally depositing the second material overlying the first

and second regions, and chemical-mechanical polishing (CMP) the second material to the level of the first material surface.

The removal of material at a localized spot using probes is possible in nonhomogeneous materials. The nonhomogeneous materials can be selected with weak interactions to the support of thin layers. A moderate pressure of the tool induces the structure formation. Scanning probe techniques are preferred in the nanometer range. For additive surface transport processes, the reservoir for the material to be deposited is in the writing probe itself. In dip-pen nanolithography, the surface-active molecules are held on the tip of a scanning force microscope. Nanostructure is formed by movement of the tip relative to the surface and by transfer of molecules via a water meniscus to the surface through the tip substrate contact point. A high affinity for the substrate surface is needed and the molecules have to be sufficiently mobile. SAMs can be formed in this fashion. One example is the octadecylthiol on a gold substrate yielding a monolayer of nanometer thickness. Spots of 15 nm diameter and structures with widths of 50 to 70 nm were realized with 16 mercaptohexaonic acid or with octadecanethiol on gold.[6]

Reshaping tools can be used to create nanopatterns. Holes with diameters below 25 nm were structured into a polymethyl methacrylate (PMMA) layer of 55 nm thickness. This process was achieved by hot embossing. The structures were prepared at 200°C with a nanolithographically fabricated tool made from a structured SiO_2 layer on a silicon wafer. The material to be shaped as a raw work piece and the shaping tool are pressed together at a temperature above the softening temperature of the material. The plastic adjusts to the shape of the shaping tool, yielding a three-dimensional replica of the tool (Fig. 4.8)[7].

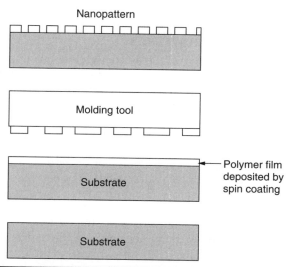

FIGURE 4.8 Hot embossing spin coated PMMA to create a nanopattern.

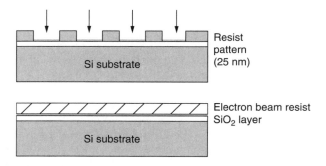

Resist
pattern
(25 nm)

Electron beam resist
SiO₂ layer

FIGURE 4.9   Electron beam fabricated tool process for fabrication of continuous nanostructures.

Nanoimprint lithography uses hot embossing procedures resulting in residues that have to be removed before further processing. Shaping techniques can be combined with multilayer resist technologies. Two or three layer systems are possible. When hot embossing is combined with lift-off, ultrasmall structures are possible. Structures with a width of 6 to 10 nm have been realized in a metal layer.

Electron beam lithography and dry etching can be used for nanoscale hot embossing (Fig. 4.9). The molding of nanopattern can be affected onto a thin film of polymer, PMMA. The reactive ion etching (RIE) process using oxygen plasma can be used to create a desired nanotopography after release of substrate from the tool.

Nanomechanical techniques include processes that incorporate a local transfer of material from a tool onto a substrate when either the tool or the substrate is prestructured. The material is transferred by nanostructured stamps through a mechanical contact onto the substrate.

4.6   Processing of Quantum Dots

When semiconductors are made small enough that quantum confinement effects become significant in all three dimensions, they are called quantum dots (QDs). Electronic and optical materials based upon QDs offer a variety of challenges to the process development engineer. Device applications of quantum wells, superlattices, and quantum wires and dots have spurred a revolution. One challenge is to form reproducible, organized arrays of QDs of any semiconductor on any surface or in a bulk matrix of glass, polymer, or ceramic. The size and distribution of QDs in a stable matrix needs to be controlled. In Type I quantum confinement devices, the confined material's band gap lies inside that of the matrix. Type II devices generally exhibit confined electrons and delocalized holes because the valence band of the confined material is below the valence band of the matrix. The Type II devices are used for infrared (IR) applications and the Type I devices are used in visible light applications. The quantum stark effect is the association of the size distribution and morphology of QDs in their matrices with the energy of optical absorption edge, control, and efficiency of electric field modulated optical absorption. Techniques are developed to enhance the uniformity, control morphology, and determine the spatial distribution of QDs in thick and thin films. These techniques include sol-gel, solid-state precipitation in glass or crystalline matrices, molecular beam epitaxy, chemical vapor deposition, and lithographic mechanisms.[8]

Coloring is affected in filter glasses by precipitation of semiconductor microcrystallines in glasses. Stark shift has been reported to be observed in semiconductor doped glass. Cuprous chloride QDs in host glasses for improved quantum confinement behavior have been found in certain borosilicate glass with alumina. The growth rate of QDs was proportional to one-third power of the heat treatment time. Precipitation kinetics could be controlled in the refractory glass. Improvement to the size distribution of QDs in glasses is possible by double annealing the glass into the nucleation and then growth regimes of precipitation. The concentration of the semiconductor plays an important role in size dispersion of QDs in glass. The lower the concentration of the dopant, the larger is the QD sizes with less dispersion. Dissolution of crystallites into an unsaturated host matrix is an alternate for precipitation of QDs. Due to the infinite solubility of SiO_2 in glass, the high temperature etch can be stopped at predetermined times leading to various sizes of the QD. QD composites and QD core-shell morphologies have been developed, leading to novel quantum optoelectronic devices.

QDs can be prepared at low temperatures by precipitation from solution by sol-gel methods. A metal acetate crystal is precipitated in the matrix and then converted to metal oxide by heating in oxygen and then to a sulphide by exposure to H_2S. Matrix plays a role in limiting crystallite size. QDs of CdSe in an organic solvent begins to convert to a hexagonal phase below 300°C. Capped CdS, QDs, with a well-defined tetrahedral morphology can be prepared with a coat of the nanocrystals formed with thioglycerine instead of with H_2S. Optically transparent and anisotropic QD thick film was formed by polymer capped QDs oriented on a quartz slide into fibers. Grains of CdS nanocrystals with size ranging from 7.4 to 26.5 nm can be prepared by a solution/growth technique on glass using controllable chemical reactions among cadmium acetate, thiourea, triethanolamine, and aqueous ammonia.

Uniformly sized QDs are possible by controlling nucleation and growth within a lithographically or electrochemically designed template. FCC packing of silica balls can be used as templates for melt/infusion of the semiconductor InSb.

4.7 Sol-Gel Processing

Sol-gel processing can be used to produce nanostructures of metal oxides. These materials may be used in explosives, propellants, and other energetic materials. Chemical reactions are conducted in solution to produce nanosized particles called sols. The sols are linked to form a three-dimensional network called a gel. The excess solvent remains in the open pores. Controlled evaporation of liquid phase results in a dense porous solid called the xerogel. On the other hand, when super-critical extraction is used, highly porous lightweight solids called aerogels are produced. A gel structure has particles and pores between them are on the nanoscale region. The University of California[9] patented a procedure where stable, inexpensive metal salts and solvents such as water and ethanol were used. The metal salt is dissolved in a solvent followed by the addition of a proton scavenger such as an epoxide. Protein scavengers react with hydrogen from the hydrated-metal and then undergo hydrolysis.

4.8 Polymer Thin Films

In applications in the semiconductor industry, polymer structures are required on length scales down to individual molecules. A "bottom-up" approach is better than a "top-down" approach in order to achieve this. A lateral resolution less than 100 nm can be created by surface instabilities and pattern formation in polymer films. Steiner[10] discussed demixing of polymer blends and pattern formation by capillary instabilities for nanostructure formation.

As a matter of generality, two different polymers are immiscible with each other. A compatible blend is one when two polymers are mixed and the property of the blend shows some improvement with commercial applications. Different from this is the requirement of molecular mixing. When two or more polymers mix with each other at a molecular level, they are considered miscible blends. Control of the morphology of phase separated polymer blends can result in nano-structures. Polymer blends can also be partially miscible.[11] Miscible polymer blends exhibit a single glass transition temperature much like a homopolymer made out of a single monomer. Immiscible poly-mer blends will exhibit the same two glass transition temperatures as their homopolymers that went into making the blend. The partially miscible blends can be expected to exhibit glass transition tempera-tures distinct from the homopolymer glass transition temperatures.

When polymers undergo phase separation in thin films, the kinetic and thermodynamic effects are expected to be pronounced. The phase separation processed can be controlled to affect desired morpholo-gies. Under suitable conditions, a film deposition process can lead to pattern replication. Demixing of polymer blends can lead to structure formation. The phase separation process can be characterized by the bimodal and spinodal curves. UCST is the upper critical solution temperature, which is the temperature above which the blend constitu-ents are completely miscible in each other in all proportions. Not found that often in other systems other than among polymers is the LCST behavior. The LCST is the lower critical solution temperature. This is the temperature above which the polymers that were miscible below this temperature now exhibit immiscibility. The bimodal and spinodal curves can be calculated by the stability criteria of Gibbs free energy.

$$\Delta G = \Delta H - T\Delta S$$

$$\Delta G \leq 0$$

$$\frac{\partial^2 G}{\partial \phi^2} > 0$$

where G is the Gibbs free energy, H is the enthalpy, S is the entropy, and ϕ is the phase composition.

When the free energy change is negative, the polymer blend can be expected to be miscible. The immiscible and miscible region is separated by the binodal curve. The spinodal curve can be found within the binodal where the curvature of the free energy curve becomes negative. When the material falls on the spinodal region of the phase diagram, spinodal decomposition can be expected to occur.

Phase morphology with a single characteristic length scale can be synthesized by quenching a partially miscible polymer blend below the critical temperature of demixing. A well-defined spinodal pattern is formed and becomes larger with passage of time. Polymer films can be made by spin coating solvent casting method. The polymers and solvent form a homogeneous mixture at first. Solvent evaporation during spin coating causes an increase in polymer phase volume, resulting in traversal to the spinodal region of the phase diagram. This can be expected to lead to polymer–polymer demixing. Characteristic phase morphology can be found in the polymer film as shown in Fig. 4.10. The polystyrene and poly vinyl pyridine are mixed in the tetrahydrafuran (THF) solvent.

Different phase separated morphologies can be found in different polymer solvent systems. The pattern formation consists of several stages. In the initial stage, phase separation results in a layered

FIGURE 4.10
Structure formation in polymer film from spin coating solvent casting.

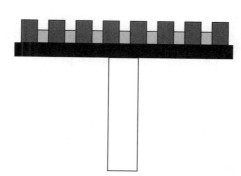

morphology of the two solvent swollen phases. As more solvent evaporated, this double layer is destabilized in two ways: (1) capillary instability of the interface and (2) surface instability. Each of the mechanisms results in different morphological length scales. Core-shell spherical domains in phase separated ternary systems have also been found. The shell thickness can be a few nanometers.

4.9   Cryogenic Ball Milling

Nanoscale titanium can be affected using cryogenic milling. Large titanium grains are subjected to severe mechanical deformation into ultrafine nanopowder, degassed, and consolidated under controlled pressure and temperature by cryogenic milling. The devices using the nanopowders exhibit improved mechanical performance characteristics.

This is a "top-down" approach for nanostructuring. Cryomilling uses a higher energy attritor-type ball-milling device in the powder metallurgy industry. This obviates subsequent heat-treatment steps. The grain size can be severely reduced and tremendous material performance improvements can be achieved. Mechanical alloying of metal powders supercooled in liquid hydrogen or nitrogen nanocrystalline materials can be used in extrusions and forgings. The raw material powder is fed into a cryogenic, attritor-type ball-milling device, processed by stirring in a medium consisting of stainless steel balls resulting in the homogenization of raw material powder. The resulting grain size that can be expected is 100 to 300 nm. The attritor speed used by Boeing[12] was 100 to 300 rpm for 8 h.

Coarse grain titanium powder material with an initial size of 50 μm is introduced into the cryomill as shown in Fig. 4.11 along with liquid nitrogen at a temperature of −196°C to form a slurry mixture. Stearic acid is also used as an additive to provide lubrication for the process. The stirring chamber is provided with a stirring rod that is coupled to a motor that can be used to control the rotational rate. The powder is allowed to come in contact with the milling medium such as the stainless steel balls shown in Fig. 4.11. The balls are moved by the stirring rod to ensure the severe mechanical deformation that is needed. At 100 to 300 rpm, sufficient grinding or milling action is achieved. Ultrafine nanoscale materials are collected through an outlet. The size reduction achieved is down to 100 to 300 nm. The next step involves degassing under a nitrogen atmosphere at a temperature of 850°F and 10^{-5} torr for 72 h. This is needed to reduce the hydrogen content in the milled powder. Then a high-pressure process consolidates the powder at 850°F and 15 KSI for 4 h. A ceracon-type nonisostaic forging process is used by Boeing to achieve the consolidation. This is then processed into fasteners using a hot or cold forming technique.

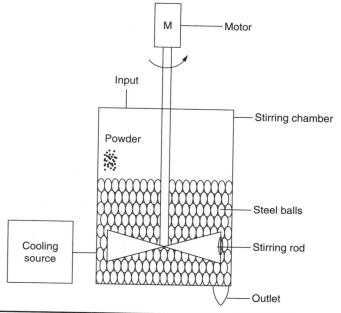

Figure 4.11   Cryogenic ball mill.

4.10   Atomic Lithography

One of the reasons for the rapid rise of modern computer hardware has been the constant miniaturization of electronic devices and integrated circuits. Photolithography has been the traditional method for IC production in the industry. When feature sizes of less than 50 nm are required, photolithographic techniques are not sufficient. In atomic lithography, conventional roles occupied by light and matter are reversed. As the light beam is patterned by a solid mask, a beam of neutral atoms is patterned by a mask of light. Long thin strips can be made using atomic lithography. A source of atoms of metals such as chromium or aluminium can bombard a substrate from an oven. Only periodic patterns were initially possible. It was necessary to develop a method to generate arbitrary patterns. The position of atom spot deposition needs to be controlled without mechanical motion. Atoms can be deposited on a substrate. An atom beam was formed.[13]

A laser beam passes to form a high intensity optical spot by interference to focus selectively on the collimated atomic beam. A spot is synthesized that can be moved to form a two-dimensional pattern of atoms on the surface of the substrate. Arbitrary two-dimensional nanostructures on a substrate can be generated using spatial modulators and lenses. The lenses and spatial light modulators are configured to selectively focus atoms in an atomic beam onto the surface.

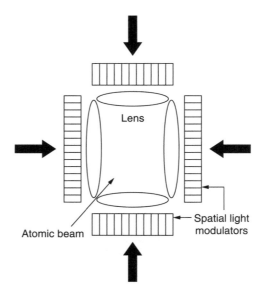

Lens

Spatial light
modulators

Atomic beam

The atomic lithographic system is shown in Fig. 4.12. Four spatial light modulators are positioned optically ahead of four lenses. The spatial light modulators are devices that have a linear array of individual light values which modulate light and can be electronically controlled to focus atom deposition. The spatial light modulators can control an optical wave front over the full range of the plane allowing the position of the atom spot deposition to be controlled precisely without mechanical motion. Spatial light modulators are available off the shelf and can be purchased from organizations such as Cambridge Research and Instrumentation, Inc., Optics Model Shape Shifter, and JenOptic.

Complete coverage of a 2π angular range is possible using lenses and spatial light modulators. The incident fields are imaged to a point source of circular waves in order to create a localized bright spot in the center of the pattern. A point source creates an outgoing wave proportional to $e^{f(kr-\omega t)}$. In order to create a spot at the center, the atomic region is illuminated with waves $e^{f(-kr-\omega t)}$ where r is the distance from the location of the bright spot. The atoms are heated in an oven and presented to the desired area by the beam. The light passes through the lenses and spatial light modulators. The spatial light modulators electronically focus atom deposition as determined by a control system so that atoms can be selectively positioned on a substrate at nonperiodic locations. In this manner, 80 nm diameter spots can be generated.

Fourier optics can be used to describe the focusing of a beam in terms of interference of the constituent plane wave components taking into account relative plane wave components and relative phase shifts due to propagation. Image synthesis can lead to direct control of the phases of each of the plane wave components.

Only one of the modulators can be used for illumination when a single segment reaches 1400 nm length and 300 nm width. Several parallel lines can be displaced relative to each other using phase offsets in a manner analogous to the spot generation shown in Fig. 4.12. Full width half maximum (FWHM) atomic line width of 300 nm can be achieved using a Gaussian beam.

4.11 Electrodeposition

Nanostructured thin film nanocomposites can be manufactured using an electrodeposition method. Metal chalcogenide nanostructured films can be made by electrodeposition. Thin films of metals and semiconductors play vital roles in advanced technologies. Improved performance of thin-film-based devices such as nanosensors, magnetic storage media, and nanoscale optical devices can be achieved through precise control of the film structure on a nanoscale. Two kinds of processes can be used for such nanostructuring. These are "dry processes" and "wet processes." Dry processes for example include vacuum evaporation, CVD, molecular beam epitaxy (MBE), and sputtering. These are conducted under vacuum and in the gas phase. Wet methods include chemical bath deposition (CBD) and electrochemical deposition (ED). Template-based synthesis is a more efficient method of preparation of nanostructured thin film nanowire arrays. A variety of nanostructures has been electrochemically deposited within porous alumina and polycarbonate templates providing arrays of nanoscale wires. Nanowires posses dimensions of approximately 20 to 250 nm and lengths of 1 to 10 mm. Materials used in preparation of nanowires include metals such as gold, silver, cobalt, copper, nickel, palladium, and platinum, and semiconductors such as cadmium sulfide (CdS), cadmium sellenide (CsSe), indium phospide (InP), gallium arsenide (GaAs), and conductive polymers such as polyacetylene.

Tulane University[14] patented an electrodeposition method to make nanostructured films within the pores of mesoporous silica by formation of a metal silica nanocomposite. The nanocomposite is annealed to strengthen the deposited metallic composition at 25 to 75 percent of the melting temperature of the deposited metal. The silica is then removed from the nanocomposite by dissolution in a suitable etching solvent such as hydrofluoric (HF) acid. Sufficient structural integrity and mechanical strength of the nanostructured film are needed.

The dimensions of nanoscale wires of metal chalcogenide thin films are controlled by the dimensions of the pores in the mesoporous silica template used to prepare the film. Nanowire structures are prepared from hexagonal templates possessing arrays of straight pores. The nanomesh is composed of a continuous, interconnected, porous nanofiber network having segments with nanoscale cross-sectional dimension. Cubic mesoporous silica templates having a network of three-dimensional interconnected pores can be used to prepare nano-mesh structures.

A mesoporous silica film is prepared by depositing a surfactant micelle-templated silica sol onto an electrically conductive substrate and calcining the deposit to provide a surfactant-free mesoporous silica film. The pore structure of the film is controlled by the choice of the surfactant and the surfactant concentration. The film is deposited on a conductive metal surface such as copper, gold, or silver to provide an electrode for electrodeposition of metal chalcogenide within the pores of the film. The film with its conductive substrate is immersed in a solution of metal ion complex and elemental chalcogen such as sulfur, selenium, or tellurium and a voltage potential is applied between the conductive substrate in contact with the meso-porous silica template and a counter electrode in the solution. The nanowires grow in a single direction, growing outward through the pores of the silica forming a nanocomposite of silica and metal chal-cogenide. The nanocomposite is later annealed and the structural integrity of the film is imparted in this step. The silica is removed by solvent etching techniques.

These nanostructured films can be used in solar cells (Fig. 4.13), catalytic membranes, and information storage media. As shown in

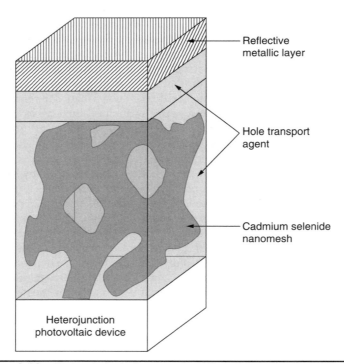

Reflective metallic layer

Hole transport agent

Cadmium selenide nanomesh

Heterojunction photovoltaic device

FIGURE 4.13   Photovoltaic device nanomesh film ccomponent.

Fig. 4.13, a self-supporting metal chalcogenide nanomesh film can be used as a photoelectronic component in the heterjunction photovoltaic device such as a solar cell. The solar cell shown in Fig. 4.13 comprises CdSe, a cadmium sellenide nanomesh film. A layer of CdSe nanomesh is in contact with an optically transparent ITO substrate. The pores of the nanomesh are filled with a hole transport agent such as a conductive polymer such as polythiopene. A reflective metallic layer is coated over the nanomesh and hole transport agent. Light entering through the ITO substrate excites the semiconductive CdSe nanomesh, generating an electron that is injected into the hole transport agent and migrates to the conductive metal layer generating an electric current. This is expected to increase the photovoltaic efficiency of solar cells from 14 percent to the order of 40 to 50 percent by use of nanostructured components. This can change the way electricity is generated for the scores of energy requirements of every man, woman, and child in the universe.

4.12 Plasma Compaction

Nanoscale particles can be formed from semiconductor compounds and thermoelectric components by using the method of plasma compaction. For example, lead tellurium (PbTe) base compounds can be used in solid-state thermoelectric cooling and electric power generation devices. The figure of merit of such devices is given by

$$Z = \frac{S^2 \sigma}{k}$$

where S is the Seebeck coefficient, σ is the electrical conductivity, and k is the thermal conductivity. Nanostructured materials can result in improvements in a figure of merit of thermoelectric device. The nanoparticles are formed from Group IV to VI of the periodic table of all elements. A solution of Group IV reagent, a Group VI reagent, and a surfactant is used during the synthesis of the nanostructure. The resultant solution along with a reducing agent is maintained at a temperature range of 20 to 360°C for a duration (1 to 50 h) sufficient to generate nanoparticles formed as binary alloys. Nanoparticles are characterized by their largest dimension in the range of 10 to 200 nm. Surfactants such as polyethylene glycol (PEG) at a concentration of 0.001 to 0.1 molar solution may be used.

A selected quantity of a surfactant is mixed in a solvent to obtain a surfactant-containing solution. Surfactants generally refer to amphiphilic molecules in which part of the molecule is hydrophilic and another part is hydrophobic. The concentration of the surfactant can be selected based on factors such as the desired average sizes and the shapes of nanoparticles generated in subsequent steps. The solvent may be polar or nonpolar. The reducing agent serves as a source of hydrogen atoms. One example of a reducing agent is sodium borohydride ($NaBH_4$). A base is also added to facilitate the reaction between Group IV and Group VI elements in order to deliver the desired nanoparticles. Nanoparticles may be expected as a reaction product after the reactant mixture is subjected to sufficient time, under the prescribed temperature and pressure. The reaction product is separated from the reactant/product mixture by centrifugation, washing, and precipitation of nanoparticles. This step is followed by drying under vacuum and packing in an inert atmosphere such as under the glove box in an argon environment.

The nanoparticles are packed in a fashion that the formation of oxide layers is inhibited. The synthesized nanoparticles are compacted at an elevated temperature and under compressive pressure to generate a thermoelectric composition.

Plasma pressure compaction apparatus has been patented by MIT[15] and is shown in Fig. 4.14. The apparatus comprises two high strength pistons that are capable of application of high compressive pressure in the range of 100 to 1000 MPa to a sample of nanoparticles that is disposed within a high strength cylinder, a current source that applies a current through the sample for heating it. The current density is in the range of 500 to 3000 A/cm^2. The temperature of the sample can be obtained by measurement of the temperature of the cylinder by optical pyrometer. The time duration of applied pressure and current is varied to achieve the desired level of compaction of nanocrystals and prevent agglomeration.

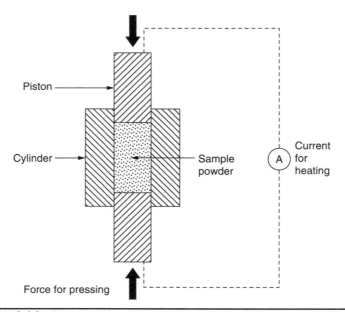

FIGURE 4.14 Plasma compaction apparatus.

This route of synthesis of nanoparticles of binary alloy nanostructures provides a high yield and throughput of the product. The reaction parameters such as temperature, surfactant concentration, and the type of solvent can be adjusted easily to vary the size and morphology of the product nanostructures.

As an example, PbTe nanocrystals were prepared by mixing 50 mg of PEG surfactant with a molecular weight of 20,000 with 50 mL of water to obtain a surfactant-containing aqueous solution. Added to the solution were 2.4 g of NaOH pellets to obtain a 1.2 molar solution. The powder and lead acetate at a concentration each of 1 mmol were added to the solution while continuously stirring it. This was followed by adding approximately 5 mL of hydrazine hydrate reducing agent to the solution and transferring the solution into a pressure vessel of 125 mL capacity. The vessel was placed in a furnace to achieve a temperature of 160°C for nearly 20 h. Later the reaction product was washed with distilled water to strip off by-products from the synthesized PbTe nanoparticles.

The PbTe nanoparticles were examined using TEM, scanning electron microscopy (SEM), and x-ray diffractometry (XRD). A high-resolution TEM image of PbTe nanocrystal is shown in Fig. 4.15. It can

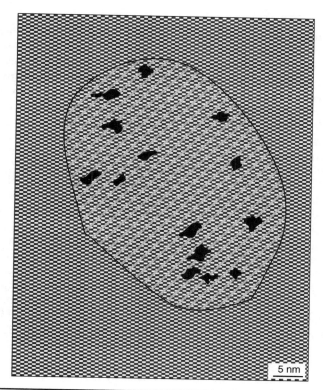

5 nm

Figure 4.15   High-resolution TEM image of nanocrystal from plasma compaction.

be seen that PbTe nanocrystals exhibit a high degree of crystallinity and they are found to be free of amorphous oxide layers surrounding their outer surfaces. This results in an improvement of the figure of merit of thermoelectric components made out of these nanocrystals. The XRD spectrum of the PbTe nanocrystals reveals that the nanocrystals are made out of FCC Bravais lattice structure with a lattice constant of 6.46 A. An average particle size of 30 nm can be calculated using the Debye-Scherrer formula used in standard materials science textbooks. In the absence of surfactant, the average particle size of the nanoparticles will increase and the crystalline quality of the nanoparticles would degrade.

4.13   Direct Write Lithography

A tip can be used to pattern a surface and prepare polymeric nanostructures. A tip is used to initiate polymerization and pattern a structure. The structure is exposed to monomer to induce polymerization at the structure. Ring opening metathesis polymerization can be carried out with the use of the tip in order to control the polymerization. The tip can be a sharp tip as used in AFM. Norborene types of monomers can be used. Biological macromolecules can also be prepared.

A technological need exists to better prepare metal, polymer, ceramic, and glassy structures at the nanoscale including preparation of patterned structures at high resolution and registration. Application areas include catalysis, sensors, and optical devices. Photolithography, electron beam lithography, and microcontact printing have been developed for generating polymer arrays. These methods do not allow for control of nanostructure in a site-specific manner.

Direct write nanolithographic printing was patented by Northwestern University.[16] Polymer brush nanostructures can be synthesized using ring opening metathesis polymerization (ROMP). Control over feature size, shape, and inter-feature distance is possible. The capability of directly delivering monomers via a tip such as in AFM's cantilever, which allows the generation of small, diverse libraries of nanoscale polymer brushes by assembly of many possible combinations of small building monomers with active functionalities, has been acquired. Edge to edge distance of less than 100 nm was found to be possible.

The polymer brush nanostructure is shown in Fig. 4.16. This is a topographic AFM image of polymer brush lines. The speed of norbornenyl thiol deposition, polymerization time, and measured average FWHM values are 0.02 mm/s, 60 min, and 480 nm, respectively. In Fig. 4.17, the graphical representation of surface initiated ROMP via dip pen (DPN) lithography printing is shown.

A nanostructure is patterned with a polymerizable compound on a substrate by first adding polymerization catalyst to the nanostructure

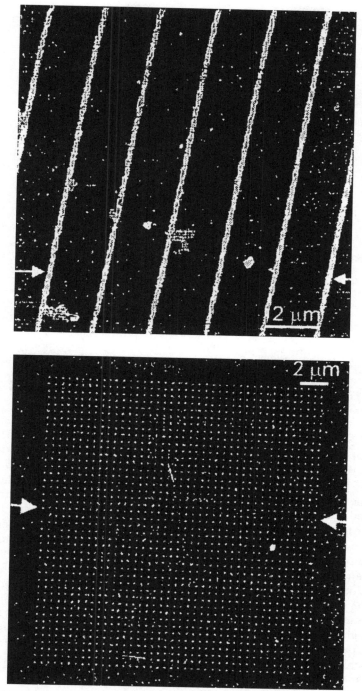

FIGURE 4.16   Polymer brush nanostructures.

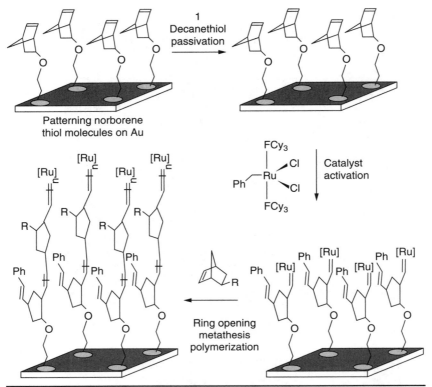

FIGURE 4.17 Surface initiated ring opening metathesis polymerization via direct write lithography.

of polymerizable compound to form a second nanostructure. This can initiate polymerization reaction. Polymerizing monomer is added to the second nanostructure to form the third nanostructure and so on. By this method, polymer brush structures can be formed where one end of the polymeric molecular structure is bonded to a surface and the other end is free from the surface extending away from the surface. The polymer is surface immobilized or end immobilized and can be anchored to the surface by covalent bonding, chemisorptions, or physiosorption. Thus, the surface can be coated or covered with a patterned polymeric structure. The surface properties can be tailored. When the substrate surface comprises a pattern, the regions between the patterns can be passivated as needed for further processing. For example, a pattern can be generated which is hydrophobic and the regions between the patterns can be made hydrophilic. Alternatively, a pattern can be generated which is hydrophilic and the regions between the patterns can be made hydrophobic. The patterned regions can comprise polymeric material, whereas the regions between the patterns can be nonpolymeric.

4.14 Nanofluids

The heat transfer in fluids can be enhanced by dispersing nanoconducting particles in the fluids. Conventional heat transfer fluids suffer from fundamental limits on heat transfer properties. Metals in solid form posses several orders of magnitude larger thermal conductivity compared with pure fluids. The thermal conductivity of copper, for instance, is 3000 times greater than that of engine oil. Low thermal conductivity of liquid is a limitation in the development of energy-efficient heat transfer fluids required in many industrial applications. Nanofluids can overcome these limitations. These are developed by suspending nanocrystalline particles in pure liquid such as water, oil, or ethylene glycol. The major dimension of the nanocrystal is less than 100 nm. The resulting nanofluids exhibit higher thermal conductivity compared with the pure fluids. Suspension properties without settling of the nanoparticles over periods longer than several days are expected. Maxwell published some theoretical work on thermal conductivity of suspensions 100 years ago. The effective thermal conductivity of suspensions containing spherical particles increases with volume fraction of the solid particles according to Maxwell's model. The thermal conductivity of suspensions increases with the ratio of the surface area to volume of the particle. Per Hamilton and Crosser's model for constant particle size, the thermal conductivity of a suspension containing larger particles more than doubled by decreasing the sphericity of the particles from a value of 1.0 to 3.0. The surface area to volume ratio is 1000 time larger for particles with 10 nm diameter compared with particles that are 10 mm in diameter. Hence, a more dramatic improvement in effective thermal conductivity is expected because of decreasing the particle size in a solution to the nanoscale.

Nanocrystalline particles were produced by direct evaporation into a low vapor pressure liquid. A cylinder under vacuum heats the substance to be vaporized in a crucible placed within the cylinder. The cylinder with the liquid is rotated to transport a thin layer of the liquid on the surface. The liquid is cooled by a cooling system to keep the liquid from increasing the vapor pressure inside the cylinder. Thioglycolic acid was used as a stabilizing agent. Agglomeration of nanoparticles needs to be prevented.

Nanocrystalline powders were prepared by gas condensation process and then subsequently dispersed in deionized water to prepare the nanofluid.[17] Nanocrystalline copper and alumina powders were produced at Argonne National Laboratory, Argonne, Illinois. Stable suspensions of commercial oxide powders in water are produced. TEM was used to characterize particle sizes and agglomeration behavior.

Another method for nanofluid preparation is the VEROS method. VEROS stands for vacuum evaporation of particles onto running oil substrate. Nanocrystalline particles are produced by evaporation into a low vapor pressure liquid. Nanocrystalline copper was evaporated into two types of pump oil. It was found that the thermal conductivity of the liquid increases dramatically as the particle volume fraction is increased.

4.15   Nanostructures by Self-Assembly of Block Copolymers

Materials with nanoscopic features have several interesting applications in the areas of low dielectric materials, catalysis, membrane separation, molecular engineering, photonics, biosubstrates, and various other fields. Synthesis methods include the use of polymers as templates, phase separation processes, and multistep templating chemical reactions. Deficiencies in these methods include complexity in execution, economically prohibitive for scale-up, and the materials formed are mechanically weak, nonuniform in structure, and of little use in application.

IBM patented a process for making materials with predefined morphology.[18] Table 4.3 provides a list of experiments and theoretical predictions for forming materials with predetermined morphology. Self-assembly of block copolymer principles is used in this approach. Let the block copolymers be characterized by volume fractions, ϕ_1 and ϕ_2, respectively. A cross-linkable polymer that is miscible with the second block copolymer is also added to the mixture. The volume fraction of the cross-linkable polymer is ϕ_3. The first and second block species have a volume fraction of ϕ_{1A} and ϕ_{2A}, respectively, in the two components of the mixture. The mixture is applied to a substrate. The assembly consists of a substrate, mixture coating/structural layer having nanostructures, and an interfacial layer without nanostructures. The thickness of the layer is 0.5 to 50 nm.

One example of a self-assembling block copolymer is polystyrene-b-polyethylene oxide (PS-b-PEO). Typically, the volume fraction of the PS block and the PEO block present in the copolymer is in a range of 0.9 to 0.1. The cross-linkable polymer used was polymethylsilsesquioxane (PMSSQ). The PMSSQ is miscible with the PEO block of the copolymer. The PS-b-PEO and PMSSQ are dissolved in toluene and 1-propoxy-2-propanol, respectively. A thin film of the mixture was deposited onto the substrate and was spin cast. A typical spin speed was 3000 rpm and falls in the range of 50 to 5000 RPM. The mixture was spin cast and annealed at a temperature of approximately 100°C for about 10 h. After annealing, the mixture was further heated. The PMSSQ crosslinks at a temperature in the range of approximately 150 to 200°C. The result of the spin casting process is the material having a predefined morphology. The spin casting process may be replaced with photochemical irradiation, thermolysis, spraying, dip coating, and doctor blading. The predetermined morphologies and block volume fractions of copolymer and cross-linkable polymer are given

Entry	ϕ_1	ϕ_2	ϕ_{1A}	ϕ_{2A}	ϕ_3	$\phi_{2A}+\phi_3$	Morphology
1	0.9	0.1	0.81	0.09	0.1	0.19	Nanospheres
2	0.9	0.1	0.63	0.07	0.3	0.37	Nanolamellae
3	0.9	0.1	0.54	0.06	0.4	0.46	Nanolamellae
4	0.9	0.1	0.27	0.03	0.7	0.73	Cylindrical nanopores
5	0.9	0.1	0.18	0.02	0.8	0.82	Spherical nanopores
6	0.9	0.1	0.09	0.01	0.9	0.91	Spherical nanopores
7	0.7	0.3	0.63	0.27	0.1	0.37	Nanolamellae
8	0.7	0.3	0.56	0.24	0.2	0.44	Nanolamellae
9	0.7	0.3	0.49	0.21	0.3	0.51	Nanolamellae
10	0.7	0.3	0.42	0.18	0.4	0.58	Nanolamellae
11	0.7	0.3	0.35	0.15	0.5	0.65	Nanolamellae
12	0.7	0.3	0.28	0.12	0.6	0.72	Cylindrical nanopores
13	0.7	0.3	0.21	0.09	0.7	0.79	Cylindrical nanopores
14	0.7	0.3	0.14	0.06	0.8	0.86	Spherical nanopores
15	0.7	0.3	0.07	0.03	0.9	0.93	Spherical nanopores
16	0.5	0.5	0.45	0.45	0.1	0.55	Nanolamellae
17	0.5	0.5	0.4	0.40	0.2	0.6	Nanolamellae
18	0.5	0.5	0.35	0.35	0.3	0.65	Nanolamellae
19	0.5	0.5	0.3	0.3	0.4	0.70	Cylindrical nanopores
20	0.5	0.5	0.25	0.25	0.5	0.75	Cylindrical nanopores
21	0.5	0.5	0.20	0.2	0.6	0.8	Cylindrical nanopores
22	0.5	0.5	0.15	0.15	0.7	0.85	Spherical nanopores
23	0.5	0.5	0.10	0.1	0.8	0.9	Spherical nanopores
24	0.5	0.5	0.05	0.05	0.9	0.95	Spherical nanopores

TABLE **4.3**   List of Experiments to Form Materials with Predetermined Morphology

Entry	ϕ_1	ϕ_2	ϕ_{1A}	ϕ_{2A}	ϕ_3	$\phi_{2A}+\phi_3$	Morphology
25	0.3	0.7	0.27	0.63	0.1	0.73	Cylindrical nanopores
26	0.3	0.7	0.24	0.56	0.2	0.76	Cylindrical nanopores
27	0.3	0.7	0.21	0.49	0.3	0.79	Cylindrical nanopores
28	0.3	0.7	0.18	0.42	0.4	0.82	Spherical nanopores
29	0.3	0.7	0.15	0.35	0.5	0.85	Spherical nanopores
30	0.3	0.7	0.12	0.28	0.6	0.88	Spherical nanopores
31	0.3	0.7	0.09	0.21	0.7	0.91	Spherical nanopores
32	0.3	0.7	0.06	0.14	0.8	0.94	Spherical nanopores
33	0.3	0.7	0.03	0.07	0.9	0.97	Spherical nanopores
34	0.1	0.9	0.09	0.81	0.1	0.91	Spherical nanopores
35	0.1	0.9	0.08	0.72	0.2	0.92	Spherical nanopores
36	0.1	0.9	0.07	0.63	0.3	0.93	Spherical nanopores
37	0.1	0.9	0.06	0.54	0.4	0.94	Spherical nanopores
38	0.1	0.9	0.05	0.45	0.5	0.95	Spherical nanopores
39	0.1	0.9	0.04	0.36	0.6	0.96	Spherical nanopores
40	0.1	0.9	0.03	0.27	0.7	0.97	Spherical nanopores
41	0.1	0.9	0.02	0.18	0.8	0.98	Spherical nanopores
42	0.1	0.9	0.01	0.09	0.9	0.99	Spherical nanopores

TABLE 4.3   List of Experiments to Form Materials with Predetermined Morphology (*Continued*)

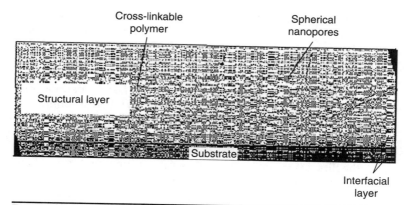

Cross-linkable polymer

Spherical nanopores

Structural layer

Substrate

Interfacial layer

FIGURE **4.18**   TEM image of material with predetermined morphology of spherical nanopores.

in Table 4.3. Morphologies that can be tailored are nanolamellae, cylindrical nanopores, spherical nanopores, and nanospheres.

A cross-sectional TEM image of a material with predetermined morphology of spherical pores is shown in Fig. 4.18. The structure consists of an interfacial layer, a structural layer, and a substrate. The substrate is in direct mechanical contact with the interfacial layer. The structural layer comprises a spherical nanopores nanostructure and essentially consists of the cross-linkable polymer. The interfacial layer lacks the spherical nanopores. The thickness of the interfacial layer is 2 to 30 nm. The structural layer thickness is in the range of 50 to 300 nm.

4.16   Pulsed Laser Deposition

Pulsed laser deposition (PLD) is a technique where pulses of laser radiation are used to evaporate material from a target, which is then deposited on a substrate in order to produce thin films with profound nanotechnological significance. Laser beams are easier to transport and manipulate and have larger dynamic ranges of delivered energy compared with ion or electron beams. PLD can be used to achieve 100 times greater deposition rates compared to other methods such as CVD, molecular beam epitaxy, plasma processing, magnetron and radio frequency (RF) sputtering, and others. However, creation of particulates during PLD prevents its usage in the semiconductor industry to generate high quality films.

Nanosecond laser pulses with low repetition rates are used to form a film with a surface free of particulates and with improved surface quality.[19] Each pulse of the laser energy is not sufficient to evaporate the target material and results in particles that get deposited on the substrate to form the desired film. Particles are eliminated from the vapor plume. The low energy of the laser pulses and optimal laser intensity are selected by thermodynamic parameters of the target resulting in improved efficiency of the evaporation process. Ultrafast laser ablation at a succession of short laser pulses at high repetition rate for heating the target and generating successive bursts of atoms and ions in a vapor plume results in deposition of thin structural films that are amorphous in character. As the laser repetition rate used increases, it is possible to reach a condition where a continuous beam of atom strikes the substrate because the spread of atom velocities in the laser-evaporated plume allows the slow atoms from one pulse to arrive at the same time as the faster atoms from later pulses. As a result, the film grows on the substrate surface from a continuous flow of atomic vapors with regulated atom flux density. This aids the creation of structured films such as epitaxial growth. In the previous method, bursts of material from successive pulses used to arrive separately at the substrate.

Typical temperature of the evaporation surface is 5000 K depending on the target material. For graphite, the average velocity of carbon atoms in the vapor flow is 2000 m/s and the expansion front moves at a typical 6000 m/s. Continuous evaporation flow of carbon atoms is found at a distance

$$d = 1.5\frac{v}{R}$$

where R is the minimum laser repetition rate. For a substrate distance of 10 cm, the minimum laser repetition rate for formation of continuous vapour flow is $R = 30$ kHz. High repetition rates much greater than R then allow the flux of atoms on the substrate surface to be fine-tuned via small variations in the laser repetition rate, permitting among other things the deposition of a single atomic layer of material. In the previous method, the number of evaporated atoms per pulse is 10 orders of magnitude higher than the modified PLD. Therefore, fine dispersion control is not possible.

The relations between laser pulse characteristics such as duration time, wavelength, laser intensity on the target surface, and pulse repetition rate R and the target material characteristics such as density, specific heat, heat conduction, and heat of vaporization from the surface allow for optimization of laser intensity for more efficient evaporation of the target. Thus,

$$I_a t_d^{1/2} = \alpha^{1/2}\rho_0\Omega$$

where I_a is the laser intensity (W/m²), α is the thermal diffusivity (m²/s) of the target material, Ω is the heat of vaporization (J/Kg), and t_d is the duration time of the pulse in seconds.

When the duration time of the pulse is reduced from nanoseconds to picoseconds or less, the evaporation process takes place in an increasingly narrow zone near the target surface. There is not sufficient time for thermal energy transport further into the target during the laser pulse. Almost all the absorbed laser energy goes into evaporation of atoms from a very thin zone near the target surface. Less absorbed laser energy is lost as waste heat. Thus, $I_a \sqrt{t_d}$ is a constant. For instance, a 10-ns pulse increased to a 100-ps pulse results in a 10 times decrease of the energy density needed to ablate the material, assuming that the laser evaporation is at the optimal regime in both cases. The advantage of using short laser pulses is the elimination of particles and droplets in the vapor flow. There is a dramatic increase in the number of atoms evaporated by a single pulse. Typically conventional PLD in optimal conditions leads to the evaporation of 10^{19} atoms/pulse, which is sufficient to generate micron-sized particles. The patented method uses very low energy pulses on 10^{11} atoms per pulse, which leads to elimination of particles in the vapor flow. This leads to better surface quality films.

A laser evaporation and deposition system is shown in Fig. 4.19. This apparatus includes a high repetition rate pulsed laser, which produces a laser beam that is directed by a series of mirrors, a focussing lens, and a laser window to a vacuum chamber. The mirrors provide a beam scanning system. The laser beam is directed to a target within the vacuum chamber. A plume of evaporated atoms is transported to a substrate, which may include a heating system. A target port and substrate port are provided for viewing the target and substrate, respectively. A vacuum pumping system and a gas filling station are connected with the vacuum chamber.

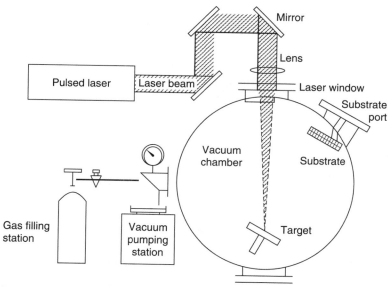

FIGURE 4.19 Laser evaporation and deposition system.

The laser deposition system includes a laser frequency conversion system, a laser beam scanning system, special telecentric focusing optics, a target manipulation system, a substrate holder and heater, a film thickness monitoring system, vacuum load locks, a pumping system with vacuum gauging, and a gas filling and flow system.

4.17 Summary

There are several kinds of nanostructures: nanowires, nanorods, nanotetrapods, nanocrystals, QDs, nanosheets, nanocylinders, nanocubes, nanograin, nanofilaments, nanolamella, nanopores, nanotrenches, nanotunnels, nanovoids, and nanoparticles. Nanostructuring methods encompass a wide range of technologies. Nanostructures can be generated either by building up from atoms using methods classified as bottom-up strategy or by diminishing the size of nanoparticles using methods grouped under a top-down strategy. Bottom-up strategies use self-assembly concepts, are cheap, more scalable, more flexible, and lead to molecular-level engineering. Top down strategies are expensive, less scalable, and inflexible.

The vacuum synthesis method of nanostructuring used sputtering of molecular ions under ultrahigh vacuums. The sputtering process is followed by the annealing process. Crystalline silicon is made to form isolated quantum wires. The gas evaporation technique is a dry process to make ultrafine metallic magnetic powders. Metal is evaporated onto a thin film under vacuum conditions. Metal atoms are allowed to impinge on the surface of the dispersing medium. Condensation of metal atoms can also be accomplished using a cooling nozzle. Nanoparticles in the size range of 50 to 4 µm can be prepared using this technique.

Triangular nanoprisms can be generated by exposure to light at different wavelengths between 400 and 700 nm. Ostwald ripening concepts are used. Edge lengths ranging from 31 to 134 nm can be prepared. Nanorods may be produced using the condensed phase synthesis method. The starting material is heated until the material vaporizes. Later the vapor is condensed. Aggregates of nanoparticles are formed. Particles are delivered by boundary layer delivery and thermophoresis-assisted deposition to form the epitaxial deposit. CNTs can also be made using this method.

Subtractive and additive fabrication methods can be used for nanostructuring operations. Lithography, etching, and galvanic fabrication processes are subtractive. A series of chemical reactions effects removal of the layer in the apertures of the mask and transfer of material into the gas phase. The lift-off process is employed in the fabrication of an IC circuit. Nanotips and nanorods can be efficiently formed using conventional CMOS processes. Patterning iridium oxide nanostructures consists of the following steps:

1. Forming a substrate with first and second regions adjacent to each other

2. Growing IrO_x structures from a continuous oxide film overlying the first region

3. Simultaneously growing IrO_x nanostructures from a continuous oxide film overlying the second region

4. Selectively etching an area of the second region

5. Lifting off overlying IrO_x structures

6. Forming a substrate with nanostructures overlying the first region. The second material can be SiO_x.

DPN, SAM, hot embossing, nanoimprint lithography, electron beam lithography, dry etching, and reactive ion etching are techniques that can be used to prepare nanostructures with 50 to 70 nm dimensions. Nanomechanical techniques include processes that include local transfer of material from a tool onto a substrate when either the tool or the substrate is prestructured.

QDs are structures where quantum confinement effects are significant. Reproducibility of organized arrays of QDs is an identified problem. Techniques such as sol-gel, solid-state precipitation, molecular beam epitaxy, CVD, and lithography were developed to enhance uniformity, control morphology, and determine spatial distribution of QDs in thick and thin films. QDs can be prepared at low temperatures by precipitation from solution by sol-gel methods. Uniformly sized QDs are affected by control of nucleation and growth of particles within a lithographically or electrochemically designed template. FCC packing of silica balls can be used as a template for melt/infusion of the semiconductor InSb.

Nanostructures of metal oxides can be prepared by sol-gel processing methods. Chemical reactions are conducted in solution to produce nanosized particles called sols. The sols are connected into a three-dimensional network called a gel. Controlled evaporation of liquid phase leads to dense porous solids called xerogel. Surface instabilities and pattern formation in polymer thin films can lead to formation of nanostructures. The control of morphology of phase-separated polymer blends can result in nanostructures. Kinetic and thermodynamic effects during phase separation can be used in preparation of nanostructures. Nanostructure can be synthesized by quenching of a partially miscible polymer blend below the critical temperature of demixing. Spin coating can be used to prepare polymer film. Pattern formation from polymer solvent systems is stage-wise. A stage of layered morphology followed by destabilization of layers by capillary instability and surface instability leads to nanostructure formation.

Cryogenic milling is a top-down approach to prepare nanoscale titanium of 100 to 300 nm size. Several mechanical deformations of large grains into ultrafine powder degassing lead to nanopowder with improved characteristics.

Atomic lithography is the method of choice to generate structures less than 50 nm dimensions. A laser beam forms a high intensity optical spot allowing formation of two-dimensionsl pattern of atoms on the surface of the substrate. Nanostructured thin film nanocomposites can be manufactured using electrodeposition method. Electrodeposition can be used to form nanostructured films within the pores of mesoporous silica. Silica is then removed from the nanocomposite by dissolution in a suitable etching solvent such as HF. Plasma compaction techniques can be used to form nanoparticles of semiconductor compounds, resulting in improvements in the figure of merit. Nanoparticles can be expected as reaction product after the reactant mixture is subject to sufficient time, under prescribed temperature and pressure. Plasma compaction apparatus may comprise two high strength pistons capable of compressive pressure in the range of 100 to 1000 MPa to a sample of nanoparticles that is disposed within a high strength cylinder. The desired level of compaction is achieved by varying the applied pressure, applied current, and time duration of the process.

In direct write lithography, a tip can be used to pattern a surface and prepare polymeric nanostructures. Polymerization is initiated by the tip and a pattern is formed. Polymer brush nanostructures can be synthesized using ROMP. Edge distances of less than 100 nm are possible and control over feature size, shape, and interfeature distance is achievable.

Nanofluids comprise nanoparticles dispersed in a suitable solvent. The surface to volume ratio is increased by 1000 times. Nanofluids are expected to have enhanced thermal conductivity comparable to that of copper. Materials with predefined morphology can be made using self-assembly of block copolymer principle. Nanospheres, nanolamellae, and nanopores both cylindrical and spherical are possible using this approach. The thickness of the interfacial layer is 2 to 30 nm. Using PLD, pulses of laser are used to evaporate the starting material and then deposited into a substrate to produce thin films with profound nanotechnological significance. The typical temperature of the evaporation surface is 5000 K, the average velocity of atoms in vapor flow is 2000 m/s, and the expansion front moves at 6000 m/s. Laser intensity is optimized for more efficient evaporation of the target.

Review Questions

1. Write a note on the vacuum synthesis method.

2. Discuss the gas-phase synthesis method.

3. Describe the sol-gel method of synthesis.

4. Discuss the direct atom manipulation method.

5. What is done differently in electrodeposition?

6. What is the condensed phase synthesis method?

7. When is the laser ablation method used?

8 What is agglomeration and how does it hamper nanofabrication?

9. Why is volume of production low in nanofabrication?

10. How is the broad size distribution of particles detrimental?

11. How is Brownian motion a concern in nanostructure?

12. Is a nanotrench a nanostructure?

13. What is the difference between nanowires and nanotetrapods?

14. What is the difference between nanocrystals and QDs?

15. Is a nanolamellae a nanostructure?

16. Nanocubes and nanograins are examples of what?

17. What are the differences between nanotrenches and nanotunnels?

18. Distinguish between a top-down approach and a bottom-up approach of nanostructuring.

19. What are the differences between wet and dry processes?

20. What does sputtering do during vacuum synthesis of nanowires?

21. Name four parameters that influence the formation of SOI and FET.

22. What is the role of etching in wave-like pattern formation during FET transistor fabrication?

23. Write short notes on the gas evaporation technique for magnetic nanopowder formation.

24. What is meant by the VEROS technique?

25. How are ferrofluids formed using the metal atom technique?

26. What is the difficulty in particle formation in the gas evaporation technique?

27. What is the range of vacuum pressure used in the gas evaporation technique?

28. How are nanoparticles stabilized during preparation of metal nanopowder?

29. What is the difference between a wet pumping system and a dry pumping system in improving the collection efficiency of nanoparticles?

30. Can the collecting medium be made of an electrostatic surface collection?

31. Name two applications of nanoclusters.

32. What are nanoprisms?

33. What is the role of Ostwald ripening during the formation of nanoprisms?

34. How are nanoprisms formed by using optical light?

35. What is the effect of a primary wavelength of light on the edge length of a nanoprism?

36. What two phases are involved in the condensed phase synthesis of a nanostructure?

37. At which step does aggregate form in the condensed phase synthesis of nanostructures?

38. How are thermophoresis and electromagnetic fields used during synthesis of a nanostructure by the condensed phase synthesis method?

39. In which method is the conversion and growth effected in the solid phase during the creation of a nanostructure?

40. Distinguish between subtractive and additive fabrication methods of creation of a nanostructure.

41. What type of fabrication process is galvanic deposition?

42. Discuss the formation of IrO_2 nanorods by the lift-off processes.

43. What is chemical-mechanical polishing?

44. What is the critical mechanism of nanostructure formation in the method of DPN?

45. SAM can be formed by what method?

46. Give the characteristic features during hot embossing of a PMMA layer.

47. What is the difference between nanoimprint lithography and DPN?

48. Distinguish between RIE and dry etching.

49. What are QDs?

50. Name two methods to control morphology of QDs in thick and thin films.

51. How can QDs be prepared at low temperatures?

52. What are sols and gels?

53. What is the difference between xerogel and aerogel?

54. Is sol-gel a chemical method or a mechanical method for nanostructure formation?

55. What are LCST and UCST?

56. When are two polymers said to be immiscible?

57. What is a binodal and spinodal?

58. What is demixing?

59. How is structure formation in a Polystyrene/Polyvinyl Pyrrolidone, PS/PVP (PS/PVP) system affected?

60. What are the roles of capillary instability and surface instability during the formation of nanostructures?

61. What are the main components of a cryogenic ball mill?

62. What is the range of operating parameters of a cryogenic ball mill?

63. What are the nanoscale feature sizes of a product from cryogenic milling?

64. What is the role of consolidation in the cryogenic milling method?

65. What is atomic lithography?

66. What are the roles of spatial light modulators and lenses in atomic lithography?

67. What are the nanoscale features achievable using the atomic lithography method?

68. Electrodeposition can be used to prepare what?

69. How does nanomesh improve the efficiency of a solar cell?

70. Why is structural integrity of nanofilm important?

71. Name two applications of nanostructured film.

72. What is figure of merit?

73. What does surfactant do during the plasma compaction method?

74. What are the main components of a plasma compaction apparatus?

75. What is the role of an optical pyrometer?

76. What are the XRD and SEM used for?

77. What is the nanoscopic feature size of nanocrystal from the plasma compaction method?

78. What are the chemical reactions involved in direct write lithography?

79. What is FWHM?

80. How does the FWHM achievable stack up for direct write lithography vs. atomic lithography?

81. What is ROMP?

82. What are polymer brush nanostructures?

83. How is an AFM used in creation of polymer brush nanostructures?

84. Why is surface to volume ratio of dispersed particles important in the design of nanofluids?

85. What is the concern of dispersed particles settling when a nanofluid is allowed to sit for several hours?

86. What is the effect of volume fraction on suspension thermal conductivity of a nanofluid?

87. How does self-assembly of block copolymers lead to predetermined nanostructures?

88. What are nanolamellae and nanospheres?

89. What are spherical nanopores and cylindrical nanopores?

90. What is the role of cross-linkable polymer in the creation of predetermined nanostructures?

91. Does the interfacial layer contain nanopores in the predetermined nanostructure with spherical nanopores?

92. What is the role of annealing in the formation of predetermined nanostructures?

93. When is doctor blading used?

94. Is the interfacial layer thickness the lowest size in the nanostructure with spherical pores?

95. How is PLD improved to prepare nanostructure film with improved surface quality?

96. What is the pulse repetitive rate?

97. How is laser intensity needed to be optimized in the PLD method?

98. What happens when pulse duration is decreased from nanosecond to picosecond?

99. What is the role of a vacuum in the PLD method?

100. What does the mirror do in the apparatus for PLD?

References

1. V. K. Smirnov and D. S. Kibalov, Methods of Formation of a Silicon Nanostructure, a Silcon Quantum Wire Array and Devices Bases Thereon, US Patent 6,274,007, 2001, Sceptre Electronics Ltd., St. Helier, NJ.
2. G. A. Pozarnsky and M. J. Fee, Process for the Manufacture of Metal Nanoparticle, US Patent 6,688,494, 2004, Cima Nanotech, Woodbury, MN.
3. C. A. Mirkin, G. S. Metraux, R. Jin, and C. Y. Cao, Methods of Controlling Nanoparticle Growth, US Patent 7,033,415, 2006, Northwesteren University, Evanston, IL.

4. D. B. Geohegan, R. D. Seals, A. A. Puretzky, and X. Fan, Condensed Phase Conversion and Growth of Nanorods Instead of from Vapor, US Patent 6,923,946, 2005, UT-Battelle, Oak Ridge, TN.
5. F. Zhang, G. M. Steeker, R. A. Barrowclift, and S. T. Hsu, Iridium Oxide Nanostructure, US Patent 7,053,403, 2006, Sharp Laboratories of America, Inc., Camas, WA.
6. S. Hong, J. Zhu, and C. A. Mirkin, Multiple ink nanolithography toward a multiple-pen nano-plotter, *Science*, 286, 523–525, 1999.
7. S. Y. Chou, P. R. Krauss, and P. J. Renstrom, Imprint of Sub-25 nm vias and trenches in polymers, *Appl. Phys. Lett.*, 67, 3114, 1995.
8. D. J. Duval and S. H. Risbud, Semiconductor quantum dots: progress, in *Processing, Handbook of Nanostructured Materials and Nanotechnology, Vol I: Synthesis and Processing*, H. S. Nalwa, Ed., Academic Press, New York, 2000.
9. T. M. Tillotsun, R. L. Simpson, L. W. Hrubesh, and A. Gash, Method for Producing Nanostructured Metal-Oxides, US Patent 6,986,818, 2006, Reagents of the University of California, Oakland, CA.
10. U. Steiner, Structure formation in polymer films: From μm to sub 100 nm length scales, in *Nanoscale Assembly—Chemical Techniques (Nanostructure Science and Technology)*, W. T. S. Huck, Ed., Springer, New York, 2005.
11. K. R. Sharma, Change in Entropy of Mixing and Two Glass Transitions for Partially Miscible Blends, 228th ACS National Meeting, Polymeric Materials: Science and Engineering, 91, 755, 2004, Philadelphia, PA.
12. S. G. Keener, Method for Preparing Ultra-Fine Submicron Grain Titanium and Titanium-Alloy Articles and Articles Prepared Thereby, US Patent 7,241,328, 2004, The Boeing Company, Chicago, IL.
13. M. E. Saffman, Atomic Lithography of Two-Dimensional Nanostructures, US Patent 6,787,759, 2004, Wisconsin Alumni Research Foundation, Madison, WI.
14. Y. Lu and D. Wang, Process for the Preparation of Metal-Containing Nanostructured Films, US Patent 7,001,669, 2006, Tulane Educational Fund, New Orleans, LA.
15. Z. Ren, G. Chen, B. Poudel, S. Kumar, W. Wang, and M. Dresselhaus, Methods for Synthesis of Semiconductor Nanocrystals and Thermoelectric Compositions, US Patent 7,255,846, 2007, Massachusetts Institute of Technology, Cambridge, MA and Boston College, Chestnut Hill, MA.
16. C. A. Mirkin, X. Liu, and S. Guo, Surface and Site-Specific Polymerization by Direct-Write Lithography, US Patent 7,326,380, 2008, Northwestern University, Evanston, IL.
17. S. U. S. Choi and J. A. Eastman, Enhanced Heat Transfer Using Nanofluids, US Patent 6,221,275, 2001, University of Chicago, Chicago, IL.
18. J. N. Cha, J. L. Hedrick, H. C. Kim, R. D. Miller, and W. Volksen, Materials Having Predefined Morphologies and Methods of Formation Thereof, US Patent 7,341,788, 2008, International Business Machines, Armonk, NY.
19. A. Rode, E. Gamaly, and B. Luther-Davies, Method of Deposition of Thin Films of Amorphous and Crystalline Microstructures Based on Ultra Fast Pulsed Laser Deposition, US Patent 6,312,768, 2001, Australian National University, Acton, Australia.

CHAPTER 5

Nanotechnology in Materials Science

<hr>

Learning Objectives

- Optical transform slides, preparation of ferrofluids, nickel nanowires, and colloidal gold experiments
- Organoclay polyamide or polyolefin nanocomposites: improved properties, improved performance, morphology and performance, characterization
- Synthesis, applications, and properties of ferrofluids
- Shape memory alloy and the martensite phase transformation
- Quantum nanowire fabrication and applications
- Liquid crystals with nanoscale dimensions
- Maghemite nanoparticles with rod shape morphology
- Synthesis and performance of nanoceramics
- Thermal barrier coatings
- Ceramic metal hybrid nanocomposites
- Three methods for synthesis of ceramic nanocomposites

<hr>

5.1   Adaption into Curricula

The high impact of nanotechnology in society can be seen with the development of light emitting diodes (LEDs), clay filled nylon nanocomposites, clay intercalated polyolefin nanocomposite, Li-ion advanced batteries, shape memory alloys, amorphous metals, and ferrofluids. The undergraduate institutions have taken steps to adapt nanotechnology into their curricula in general. In this chapter, we focus on the change in curricula in the introductory course of materials science and engineering. This has been accomplished in a number of universities including Beloit College, Beloit, Wisconsin, Christian Brothers University, Memphis, Tennessee, and Lawrence University, Appleton, Wisconsin.[1]

X-ray diffraction was illustrated as an important tool to characterize nanoscale materials in lectures and laboratories by using optical transform slides. Laser printer written, photographically reduced patterns on 35-mm slides contain a variety of arrays that mimic the packing of atoms in common metal and mineral structures. As the spacing of dots on the slide is similar to the wavelength of visible light, diffraction occurs when coherent light from a hand-held laser is passed through the slide. Students measure the feature spacing on the slide using a plastic ruler and hand lens and then compare the value with that obtained in a simple diffraction experiment using the Fraunhofer equation. The inquiry-based approach allows students to explore x-ray diffraction and develop an increased appreciation of the manner in which structural information is acquired from diffraction data.

Another popular experiment with students in the nanostructuring laboratory is the preparation of ferrofluids. Particles of magnetite are suspended in water with the aid of a surfactant. Since the ferromagnetic mineral magnetite contains iron atoms in both the divalent and trivalent oxidation states, the ratio of Fe(3+) and Fe(2+) ions utilized in the synthesis needs to be controlled in order to keep the final product, ferromagnetic.

Agglomeration of spheres is obviated by use of an ionic surfactant such as tetramethylammonium hydroxide. The surfactant ions coat the surface of the magnetite nanoparticles and electrostatically isolate the particles from one another. This kind of experiment, which is based upon concepts, is used to illustrate the principles of oxidation state, stochiometry, crystal structure, magnetism, and surfactant chemistry. Ferrofluids are used in manufacture of computer hard disk drives and loudspeakers, and in research on magnetically controlled drug delivery.

Nickel nanowires can be prepared and characterized using a template synthesis technique. Nickel is electrodeposited into the 200-nm diameter pores of an alumina filtration membrane that is readily available off the shelf by use of a nickel salt solution and an AA battery. Nanowires that are 200 nm in diameter and 50 μm in length are liberated from the membrane by dissolving both the silver cathode and the alumina template itself. Suspensions of nanowires were dispersed on a microscopic slide and observed under an optical microscope with a 10× or 50× lens. The alignment and movement of the magnetic nanowires are controlled using common magnets. Nanowires can be analyzed using SEM and powder x-ray diffraction techniques.

Colloidal gold can be synthesized by reaction of sodium citrate and hydrogen tetrachloroaurate. Sodium citrate reduces the Au ions to form a reddish colloidal suspension of gold nanoparticles. A laser pointer is shined through the solution to identify the presence of gold colloid. The laser light is scattered by the gold colloid in certain orientations allowing the observation of the laser beam in the suspension. Thus, the Tyndall effect is demonstrated. The laser beam can be made to gradually disappear and reappear in the solution by simple rotation about the axis of a polarized laser pointer beam. The differences in scattering of the polarized beam by the colloid as the polarization direction is altered are tapped into. This experiment can be used to illustrate redox chemistry, colloids, electromagnetic radiation, and polarization effects.

Students can construct a microfluidic device that filters nanoparticles from solution. A nylon membrane formed by interfacial polycondensation reaction between two immiscible solutions, diaminohexane and sabacoyl chloride, is used. The solutions are brought into contact at an interface within a microfluidic device. The interface where this reaction occurs is stabilized by the surface treatment of the device substrate with a self-assembled monolayer to create neighboring hydrophilic and hydrophobic regions. The membrane forms at the virtual wall where these regions intersect and allows the device to be used as a filter capable of separating gold nanoparticles from an aqueous solution. Several laboratory periods have to be allocated to complete the device and conduct analyses on its components. This experiment can be used to illustrate AFM, SAM deposition, and microfluidic technology.

5.2 Polymer Nanocomposites

In the area of polymer technology prior to the emergence of nanotechnology, nanoscale dimensions were seen in some areas. These are as follows:

1. Phase separated polymer blends often feature nanoscale phase dimensions.

2. Block copolymer domain morphology is often at the nanoscopic level.

3. Asymmetric membranes often have a nanoscale void structure

4. Mini emulsion particles are below 100 nm.

5. Interfacial phenomena in blends and composites involve nanoscale dimensions.

6. Carbon black reinforcement of elastomers, colloidal silica modification, and asbestos-nanoscale fiber diameter reinforcement are subjects that have been investigated for decades.

Recent developments in polymer nanotechnology include the exfoliated clay nanocomposites, CNTs, carbon nanofibers, exfoliated graphite, nanocrystalline metals, and a host of other filler modified composite materials. Polymer matrix-based nanocomposites with exfoliated clay are discussed in this section. Performance enhancements using such materials include increased barrier properties, flammability resistance, electrical/electronic properties, membrane properties, and polymer blend compatibilization. There is a synergistic advantage of nanostructuring that can be called the nanoeffect. It was found that exfoliated clays could yield significant mechanical property advantages as a modification of polymeric systems. Crystallization rate and degree of crystallinity can be influenced by crystallization in confined spaces. The dimensions available for spherulitic growth are confined such that primary nuclei are not present for heterogeneous crystallization and thus homogeneous crystallization results. This occurs in the limit of n in the Avarami equation approaching 1 and often leads to reduced crystallization rate, degree of crystallinity, and melting point. This was found by Loo[2] in phase separated block copolymers. Confined crystallization of linear polyethylene in nanoporous alumina showed homogeneous nucleation with pore diameters of 62 to 110 nm and heterogeneous nucleation for 15- to 48-nm pores. When nanoparticle is attached to the polymer matrix, similarities to confined crystallization exist as well as nucleation effects and disruption of attainable spherulite size. Nucleation of crystallization can be seen by the onset of temperature of crystallization, the half time of crystallization seen in a number of nanocomposites[3] such as nanoclay-poly-ε-caprolactone, polyamide-66-nanoclay, polylactide-nanoclay, polyamide-6-nanoclay, polyamide 66-multi-walled CNT, polyester-nanoclay, and polybutylene terepthalate-nanoclay.

The clay called montmorillonite consists of platelets with an inner octahedral layer sandwiched between two silicate tetrahedral layers. The octahedral layer can be viewed as an aluminum oxide sheet with some Al atoms replaced with Mg atoms. The difference in valencies of Al and Mg creates negative charges distributed within the plane of the platelets that are balanced by positive counterions, typically sodium ions located between the platelets or in the galleries. In its natural state, the clay exists as stacks of many platelets. Hydration of the sodium ions causes the galleries to expand and the clay to swell. The sodium ions can be exchanged with organic cations with the platelets completely dispersed in water. The organic cations can be those from an ammonium salt leading to the formation of an organoclay. The ammonium cation may have hydrocarbon tails and other groups attached and can be called a surfactant due to its amphiphilic nature. The extent of the negative charge of the clay is characterized by the cation exchange capacity. The XRD lattice parameter of dry sodium montmorillonite is 0.96 nm and the platelet is approximately 0.94 nm thick. The gallery expands when the sodium is replaced with larger organic surfactants and the lattice parameter may increase by two- to threefold. The lateral dimensions of the platelet are not well defined. The thickness of montmorillonite platelets is a well-defined crystallographic dimension. They depend on the mechanism of platelet growth from solution in the geological formation process.

Nanocomposites can be formed from organoclays and *in situ* solution and emulsion polymerization. Melt processing is of increased interest. Clays are exfoliated and are completely separated by polymers in-between them. This desired morphology is achieved to varying degrees in practice.

Three types of morphologies are possible during the preparation of clay nanocomposites: (1) immiscible nanocomposites, (2) intercalated nanocomposites, and (3) exfoliated nanocomposites. Representative schematics describing the three different morphologies are shown in Figs. 5.1 through 5.3. As shown in Fig. 5.1, for the case of immiscible morphology the organoclay platelets exist in particles. The particles are comprised of tactoids or aggregates of tactoids, more or less as they were in the organoclay powder, i.e., no separation of platelets. The wide-angle x-ray scattering (WAXS) scan of the nanocomposite is expected to appear essentially the same as that obtained for the organoclay powder. There is no shift in the x-ray lattice parameter reading. Such scans are made over a low range of angles, 2θ, such that any peaks that form a crystalline polymer matrix are not seen because they occur at higher angles. For completely exfoliated organoclay, no wide-angle x-ray peak is expected

FIGURE 5.1
Morphology of immiscible nanocomposites.

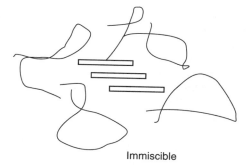

Immiscible

FIGURE 5.2
Morphology of intercalated nanocomposites.

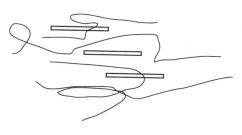

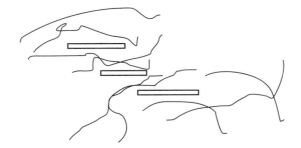

Figure 5.3   Morphology of exfoliated nanocomposites.

for the material because there is no regular spacing of the platelets and the distances between platelets would be larger than the WAXS can detect.

Often WAXS patterns of polymer nanocomposites reveal a peak similar to the peak of the organoclay but shifted to lower 2θ or larger lattice parameter spacing. The peak height indicates that the platelets are not exfoliated and its location indicates that the gallery has expanded and it is often implied that polymer chains have entered or have been intercalated in the gallery. Placing polymer chains in confined spaces involves a significant entropy penalty that must be driven by an energetic attraction between the polymer and organoclay. The gallery expansion in some cases may be caused by intercalation of oligomers. Intercalation (Fig. 5.2) may be a prelude to exfoliation. Exfoliation morphology (Fig. 5.3) may be achieved by melt processing.

There is increased interest in relating morphological structure to the performance properties of nanocomposites. WAXS analyses are frequently used. The interpretation of WAXS spectra needs to be fine-tuned. The exfoliated and immiscible morphologies need to be picked up by the WAXS spectra. Shifts in the peaks of the organoclay need to be interpreted with respect to the morphology of the nanocomposite. Small-angle x-ray scattering (SAXS) may be more quantitative and informative. Some investigators use solid state nuclear magnetic resonance (NMR) and neutron scattering. TEM is used to visualize the morphology of nanocomposites. The downside of TEM is that it is confined to a small region of the material. More scans may be needed at different magnifications so that it becomes a representation of the material considered. No staining is needed as the organoclay is sufficiently different in characteristics compared with that of the polymer matrix. In nylon 6 nanocomposites, ~1 nm thick clay platelets can be seen as dark lines when the microtone cut is perpendicular to the platelets. Image analysis can be used to quantitate the distribution of platelet lengths. In order to draw a conclusion from the analyses, hundreds of particles need to be studied. Aspect ratio distributions of the platelets can be obtained. Often in practice, the exfoliation is never complete.

The addition of fillers to polymers is to obtain an increase in the modulus or stiffness by reinforcement mechanisms. Stiffness of the material can be increased by proper dispersion of aligned clay platelets. This can be seen by the increase in the tensile modulus of the material with the addition of the filler to the product. An increase of the modulus by a factor of two is accomplished using montmorillonite with one-third of the material needed for when glass fibers are used and one-fourth of the material when talc is used with thermoplastic polyolefins (TPOs). Addition of montmorillonite to a nylon 6 matrix can result in increases of 100°C in heat distortion temperature (HDT). Depending on the level of adhesion between the dispersed phase and continuous phase, the fillers can also increase strength as well as modulus. For glass-fiber composites, chemical bonding is used to increase the interfacial adhesion. Addition of fillers results in a decrease in ductility of the material. Melt rheological properties of polymers can be drastically modified upon addition of fillers in the low shear rate region. Clay particles serve as effective nucleating agents that greatly change the crystalline morphology and crystal type for polymers like nylon 6 or polypropylene (PP).

Nanocomposites, like plastics, have a high coefficient of thermal expansion and are unlike metals, which have a low coefficient of thermal expansion. Addition of fillers can result in lowering the coefficient of thermal expansion (CTE). This can become an increased concern in automotive applications. The barrier properties of polymers can be significantly altered by inclusion of inorganic platelets with sufficient aspect ratio in order to alter the diffusion path of penetrant molecules. Exfoliated clay modified polyethylene terepthalate (PTE) has been investigated for offering improved barrier properties. *In situ* polymerized PET-exfoliated clay composites were noted to show a factor of two reduction in barrier property at 1 percent clay. A 10- to 15-fold reduction in oxygen permeability with 1 to 5 wt percent clay has been reported for PET-clay nanocomposites. The permselectivity improvement needed for better membrane separation efficiency is achieved by molecular sieve inclusions in a polymer film. Increased flammability resistance can be achieved by nanoplatelet/nanofiber modification of polymeric matrices. With nanofiller incorporation, the maximum heat release rate is reduced. Flame propagation to adjacent areas is minimized.

Addition of nanoparticles results in less coalescence of particles during melt processing, resulting in improved compatibilization. For example, exfoliated clay compatibilization such as in a polycarbonate/polymethylmethacrylate (PC/PMMA) system, polyphenylene oxide/polyamide (PPO/PA), polyamide/ethylene propylene rubber (PA/EPDM), polystyrene/polymethylmethacrylate (PS/PMMA), or polyvinyl fluoride/polyamide (PVF/PA) 6 blends is achieved by lowering the interfacial tension between the two phases that are phase separated.

There are a number of biomedical applications for polymer nanocomposites. Electrospinning for producing bioresorbable nanofiber scaffolds is an active pursuit of scientists for tissue engineering applications. This is an example of a nanocomposite as the resultant scaffold allows for cell growth yielding a unique composite system. Poly lactic acid/exfoliated montmorillonite clay/salt solutions were electrospun followed by salt leaching/gas forming. The resultant scaffold structure contained both nanopores and micropores, offering a combination of cell growth and blood vessel invasion microdimensions along with nanodimensions for nutrient and metabolic waste transport. Electrically conducting nanofibers based on conjugated polymers for regeneration of nerve growth in a biological living system is another example. Antimicrobial/biocidal activity can be achieved by introducing nanoparticle Ag, salts, and oxide of silver into polymer matrices. Hydroxyapetite-based polymer nanocomposites are used in bone repair and implantation. Drug delivery applications find a use for polymer nanocomposites. Nanoparticles can result in more controlled release, reduced swelling, and improved mechanical integrity.

Some commercial applications for polymer nanocomposites include: Toyota's nylon/clay nanocomposite timing belt cover; General Motors' TPO/organoclay nanocomposite for exterior trim of vans; Babolat's epoxy/CNT nanocomposite for tennis rackets; nitro Hybtonite's epoxy/CNT nanocomposite for hockey sticks; Inmat's polyisobutylene/exfoliated clay nanocomposite for tennis balls; SBR/carbon black nanocomposite for tires; Hyperions's polymer/MWNT nanocomposite for electrostatic dissipation; Curad's polymer/silver nanocomposite for bandages; Nanocor's nylon/clay nanocomposite for beverage containers; Pirelli's styrene-butadiene (SBR)/dispersant nanocomposite for winter tires; and Ube's polyamide/clay nanocomposite for auto fuel systems.

5.3   Ferrofluids

Aqueous magnetic ferrofluid compositions are used in magnetic imaging and printing with dry and liquid developer compositions, electrophotography, xerographic imaging, and printing. Ferrofluid possess unique electrical, optical, magnetic, and chemical properties because of their nanoscale dimensions.

In the presence of a magnetic field, ferrofluids polarize rapidly in great measure. It can be viewed as a colloid. The dispersed particles are nanoscale magnetic particles. These particles may be ferromagnetic or ferrimagnetic .

One limitation to the preparation and use of ferrofluid is the tendency of the initially small nanoparticles to aggregate into larger masses to reduce the energy associated with the high surface area to volume ratio of the nanoparticles. The dispersant is made small by milling and other top down subtractive operations. The particles may be coated with a surfactant to prevent them from clumping together.

Ferrofluids are different from magnetorheological (MR) fluids. Particles in ferrofluids are nanoscale and those in MR fluids are microscale in dimension. Xerox[4] has patented a process to prepare ferrofluids with a high magnetization moment. Fe_2O_3 nanoparticles are milled down to less than 10 nm and an aqueous liquid is added to form a colloidal suspension. The suspension is treated by centrifugation. The residue from the centrifuge is exposed to a magnetic field. The magnetic field produces a magnetic component and a nonmagnetic sludge component. The magnetization of the ferrofluid was found to be as high as 20 to 40 emu/gm. The high magnetization ferrofluid is treated with a silanating reagent such as dichlorodimethylsilane. These react with the functional groups on the surface of the nanoparticles, forming a hydrophobic coating on the particles. They may otherwise undergo hydrolysis. Alternatively, they may adsorb on the surface of the nanoparticles. In any case, the surface properties of the dispersant are altered. The hydrogel layer that forms on the surface of nanoparticles is influential in determining the shelf life stability of ferrofluids. It prevents agglomeration, precipitation, and aggregation of the particles. Ferrofluids exhibit a shelf life of 3 years at ambient to elevated temperatures. Ferrofluids exhibit a high spike forming property. The composition of the ferrofluid is 20 to 50 wt percent Fe_2O_3 nanoparticles, 50 to 80 wt percent ion exchange resin, and 50 to 90 wt percent aqueous phase. The aqueous ferrofluids are stable in the pH range of 4 to 12. Ferrofluids are more viscous compared to water. They can be used as a liquid ink. When oversilanated, ferrofluids can be used as a nanocompass.

Ferrofluids have a number of applications: liquid seals around the spinning drive shafts in hard disks, high frequency speaker drivers as tweeters, friction reducing agent, radar absorbent material, spacecraft's attitude control system, optical applications due to their reflective properties, cancer detection, removal of tumors, heat transfer, and thermomagnetic convection.

5.4 Shape Memory Alloy

An interesting nanotechnology example is the advances made in the area of shape memory alloy (SMA). SMA is a metal-based alloy and has a reversible solid-state transformation called the martensitic transformation. These materials exhibit the shape memory effect and superplasticity not found in conventional metals and alloys. They can be ferrous or nonferrous martensites. The alloys are based on iron, copper, and nickel-titanium. Approximately 48 to 52 wt percent nickel is present in nickel-titanium SMA. The transformation temperature to the martensitic phase with a monoclinic lattice structure may be sensitive to the nickel content. Transformation temperature ranges between –40 and 100°C. Tertiary elements may also be present in the alloy. Superelastic behavior is an isothermal event characterized by evidence of pseudo-elastic behavior until a critical strain is reached. In SMA materials, the superelastic effect is the result of a stress-induced martensitic phase transformation process. Upon releasing the stress, the reverse phase transformation occurs, which may result in the recovery of large strains and restoration of original dimensions. Further stressing beyond the strain level that can be accommodated by stress-induced martensitic phase transformation results in deformation that mimics conventional plastic deformation. The elastic limit of conventional metals is less than 0.2 percent. The large recoverable strain of several percent in superelastic SMA gives rise to the characterization of superelasticity.

The transition from martensite phase to the austenite phase is dependent only on temperature and stress, not on time. SMA materials can be one-way or two-way shape memory materials. When SMA is in its cold state, the metal can be bent or stretched into a host of new shapes and will hold that shape until it is heated above the transition temperature. Upon heating, the shape changes back to its original shape. When the metal cools again, it will hold the hot shape unless it is deformed another time. In two-way shape memory effect, the material remembers two different shapes—one at low temperatures and another at high temperatures.

The pseudo-elastic properties of SMA are used in commercial applications. Frames for reading glasses are made of SMA. They can deform when subject to high temperature but will return to the original shape when the stress is removed. FSMAs are ferromagnetic shape memory alloys. When subjected to strong magnetic fields, they change shape. SMA materials are manufactured by casting, vacuum arc melting, or induction melting.

Applications of SMA materials include: adhesive between two dissimilar metals, shape memory coupling for oil pipe lines, vascular stents, dental braces, Nitinol wires to straighten teeth, endodontics, materials in robotics, and eyeglass frames.

5.5   Nanowires

There is increased interest in synthesizing nanowires. Photoemission of semiconductor nanowires, GMR of multilayered nanowires, magnetic behavior of nickel and cobalt nanowires, and current-voltage response of platinum nanowires are examples of interesting applications of nanowires.

Nanowires have diameters in the range of 1 to 500 nm. They are solid materials but have an amorphous structure, a graphite-like structure, or a herringbone structure. The nanowires are periodic about their axis. Nanowires are made out of metals or semiconductor materials.

Nanowires with less than 100 nm diameter display quantum conduction phenomena such as the electron wave interference effect and survival of phase information of conduction electrons. One method of making quantum nanowires involves a microlithographic process followed by metalorganic chemical vapor deposition (MOCVD). Either one nanowire or a row of gallium arsenide (GaAs) quantum wires embedded in a bulk aluminum arsenide (AlAs) substrate can be made.

In another process, the pores or nanochannels in a substrate are filled with the material of interest. Difficulty was experienced in drawing lengthy nanowires because as the pore diameters become small, the pores tend to branch and merge and the desired material could not fill the pores. Yet another method is by pyrolysis of metalorganic precursor for self-assembly of CNTs.

Honda patented a process for preparing metal nanowires and nanostructures.[5] A substrate is provided, a metalorganic layer is deposited on the substrate, and heat is supplied to form the nanowires on the substrate (Fig. 5.4). The substrate can be made out of oxides of silicon, aluminum, magnesium or glass mica, Teflon, ceramics, plastic, or quartz. Metalorganic material can be deposited on the substrate as a thin film and heated under air to form the metal nanowires. An example of a metalorganic material is iron phthalocyanine. After depositing the precursor film, the masking layer is removed from the substrate. The metalorganic film remaining on the substrate is then pyrolyzed to form metal nanowires.

Bismuth is an interesting model system to examine low-dimensional physical phenomena due to its large anisotropy, high carrier mobility, and light effective masses. It is a semimetal with small band overlap energy. Fabrication of bismuth nanowires can lead to the observation of dependence of electronic properties on the wire diameter. Below a certain wire diameter, it becomes a semiconductor depending on the

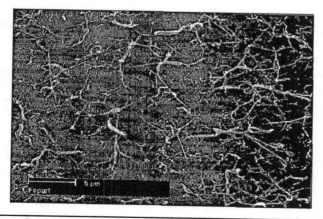

FIGURE 5.4 Iron nanowires on silicon oxide substrate.

temperature, crystalline axis of the nanowire, and the amount of antimony that is added to the bismuth. Bismuth crystals are semimetallic with an equal number of electrons and holes. This makes it undesirable for thermoelectricity. The wire diameter where the semimetal semiconductor was found to occur was 40 to 50 nm depending on the crystalline orientation of the nanowires. Arrays of hexagonally packed parallel bismuth nanowires with diameter range of 7 to 110 nm and length of 25 to 65 μm have been prepared and the synthesis methods have been reviewed by Dresselhaus et al.[6] Nanowires are embedded in a dielectric matrix of anodic alumina, which because of its array of parallel nanochannels is used as a template for preparing Bi nanowires. Materials science principles serve to confine the bismuth to these nanochannels because the bismuth does not diffuse into the anodic alumina matrix, which forms an excellent barrier material. Bismuth nanowires are crystalline and oriented with a common crystallographic direction along the wire axis. This can be picked up by standard XRD techniques. As the band structure of bismuth is highly anisotropic, the transport properties of bismuth nanowires are expected to be dependent on the crystallographic orientation along the wire axes. Crystalline state is maintained in bismuth nanowires. Therefore, many of the bulk properties of bismuth may be applicable in nanowire morphology.

An idealization is a one-dimensional quantum wire where the carriers are confined inside a cylindrical potential well bounded by a barrier of infinite potential height. An extension of this approach provides a reasonable approximation for a bismuth nanowire embedded in an alumina template in view of the large band gap of the anodic alumina template, which provides excellent carrier confinement for the embedded quantum wires. Quantum confinement effects in bismuth nanowires are more prominent than for other wires with the same diameter. Such a model can be used to predict the wire diameter for transition from semimetal to semiconductor as a function of crystallographic orientation and direction. Measurements of the transport properties of these arrays of bismuth nanowires at different wire diameter, different temperatures, and magnetic field were obtained. This demonstrated the semimetal–semiconductor transition. It showed the existence of ballistic transport at low temperatures. Reasonable agreement between theory and experiment was found. Bismuth can be alloyed with antimony (Sb) to tweak the material to the desired property needed. Sb alloying can lead to improved performance of p-type bismuth in both bulk and nanowire forms. This can be better understood by examining the phase diagram of bismuth nanowires as a function of wire diameter and Sb concentration. The semimetal and the indirect semiconductor states in the low antimony content region have the highest valence band extremum at the T-point, while those states in the higher Sb content region have their H point hole pockets at the highest energy. Along the solid lines in the phase diagram, the extrema of the first subband of carrier pockets at two different points in the Brillouin zone coalesce in energy. Thus, nanostructures can be designed with superior thermo-electric properties.

5.6 Liquid Crystals

Liquid crystals exist in a state of matter that can be considered to be in between that of solids and liquids. The molecules can move inde-pendently as in the case of a liquid state but are well organized as is typical of crystalline solids. Liquid crystals are ordered at the nanoscale but lack long-range order. The intermolecular forces that determine the interactions at the nanoscale are influential in the occurrence of macroscopic properties that are observed in fats and oils and the color changes seen in a liquid crystal. Cholesteric phase liquid crystals[7] contain molecules that are aligned in layers rotated with respect to each other. Stacks of layers are rotated with respect to one another similar to a spiral staircase or screw threads. The rotation angle from one layer to the next increases with temperature so the pitch or distance between layers with the same orientation decreases with temperature. When the pitch is about the same size as the wave-length of light (400 nm) the liquid crystal diffracts light. The wave-length of the reflected color is proportional to the pitch and will change the temperature and viewing angle. As the temperature of a liquid crystal is increased, all the colors of the rainbow will be observed from red to orange to yellow to green to blue.

5.7 Amorphous Metals

Korea Institute of Science and Technology[8] patented a synthesis method to prepare maghemite (γ-Fe_2O_3) nanoparticles that can be used as a super high-density magnetic recording substance due to its shape anisotropy and magnetic characteristics. The world semicon-ductor market and telematique industry require materials suitable for increased memory, miniaturization, and high integration of elec-tronic materials. The recording density of a commercial magnetite-recording medium has been increased steadily every year. Storage density of a magnetic recording medium is inversely proportional to the size of a storing bit. When a particle size is not greater than a threshold value of say ~10 nm, the inherent characteristics of a sub-stance are instantly changed into super paramagnetic and are hence inappropriate for use as a magnetic recording medium. This has been observed in Co, Ni, Ge or spherical magnetite, maghemite, and other ferrite particles. There have been attempts in the industry to improve crystal magnetic anisotropy in order to maintain the cohesive forces concurrently achieving reduction in particle size of a substance. In order to change the inherent paramagnetism of the spherical FePt into ferromagnetism, a heat treatment process at temperatures greater than 550°C was found necessary with the oxidation occurring in the atmosphere. Expensive platinum has to be used. Nanoscale cobalt particles with rod-shaped morphology have exhibited ferromagnetism due to shape magnetic anisotropy and are arranged parallel to each other. It has been considered as a super high-density magnetic recording substance of the future. Nanoparticles with spherical morphology have been found to exhibit superparamagnetism.

Maghemite nanoparticles with rod-shaped morphology with diameters of 0.25 μm and lengths in the ratio of 6 to the diameter find applications as high-density magnetic recording substances. The synthesized maghemite rods are converted to magnetite with the same rod morphology. Nanoparticles with less than 10 nm exhibit superparamagnetism. Particles tend to agglomerate and investigators found it difficult to synthesize particles less than 30 nm.

A flow chart of the patented process to produce magnetic nanoparticles less than 10 nm with interesting properties is shown in Fig. 5.5. A reverse micelle solution is prepared with distilled water, surfactant, and a solvent. This is added to a metallic salt, precipitating gel-type amorphous metal oxide particles. The gel-type amorphous iron oxide particles are subjected to a polar solvent wash and the molar ratio of

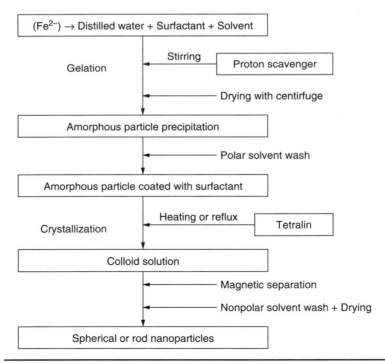

Figure 5.5 Flow chart for synthesis of metal oxide nanoparticles.

the iron oxide and surfactant is changed. The iron oxide particles are crystallized by heating or refluxing after dispersion of the gel-type amorphous metal oxide particles in a nonpolar solvent with a high boiling point. By changing the molar ratio of the distilled water to the metallic salt, the particle size of the particles in the product can be controlled. With all other variables held constant, the ratio of distilled water to surfactant can be varied in a reverse micellar solution. As the amount of distilled water is increased, the size of reverse micelle is increased and the size of gel-type amorphous iron oxide formed in each reverse micelle and the size of nanoparticle finally crystallized can be increased. This has to be within the critical micelle concentration (CMC).

The hydrophilic functional group selected for use as a surfactant is COOH or NH_2. The shape of nanoparticles obtained can be controlled by coating the particles with surfactants. The solvent used in the process is dibenzylether or diphenylether. As the reverse micelle gets homogeneously dispersed, the iron oxide particle size distribution can be narrowed further. The proton scavenger used is selected from ethylene oxide, epoxy pentane, epichlorohydrin, etc. Epoxy compounds are derived from the gel of amorphous iron oxide by performing the gelation reaction slowly and by capture of a proton from a hydrate of trivalent metal salt. Catalysts may be used to facilitate better yield and throughput. Polar solvents such as methanol, ethanol, or propanol are used to wash the gel-type amorphous iron oxide particles. The negative ions are eliminated upon washing. The number of washing times can be used to control the shape anisotropy of the crystallized iron oxide particles. Crystallization is induced by heating or reflux. When the washing time limit is exceeded, aggregation and growth of particles occur. The limit is approximately 2 to 6 times. Tetralin is used as a nonpolar solvent. It is possible to crystallize amorphous iron oxide as a maghemite phase, a hematite phase, and a maghemite/hematite mixed phase. Temperature and drying conditions can be changed to change the composition of the product. As the reflux times increase to more than 10 h, the magnetism of nanoparticles increases to approximately 10.3 emu/gm. At long times, the crystallinity of the particles increased.

Rod-shaped maghemite or hematite with a particle diameter of 2 to 10 nm and length/diameter (L/D) ratio not less than 1 and not greater than 10 can be obtained.

5.8 Nanoceramics

Ceramic materials are used in a wide range of applications. They possess excellent heat, corrosion, and abrasion resistance, and exhibit unique electrical, optical, and magnetic properties. Examples of ceramic materials include metal nitride, boride, and carbide. Nanocrystalline ceramics with 1 to 500 nm dimension have been reported in the literature. In 2002, the total U.S. market for advanced ceramic powders, including nanopowders, was approximately $1.6 billion. The projected growth rate of the industry was 7.3 percent. The total consumption of nanoceramic powders was approximately $154 million and the growth rate of the industry was 9.3 percent. The field of materials science and engineering is being revolutionized by the development of advanced ceramics.

The methods used to synthesize nanocrystalline ceramics can be classified into three categories: (1) chemical processing, (2) thermal processing, and (3) mechanical processing. Fine powders are obtained by milling procedures such as using ball mill from large particulates in a mechanical processing method. Fragmented powders that result from mechanical methods can yield different shapes and sizes that are not suitable for better performing applications. In the chemical processing method, precipitation of particles with different shapes and sizes is effected from organometallics. Sol-gel or solution-based gelation techniques for ceramic production are also increasingly reported in the literature. These approaches involve higher cost of raw materials and capital equipment.

In thermal processing, nanoparticles are collected through rapid cooling of a supersaturated vapor. Metallic raw material is evaporated into a chamber, heated to very high temperatures, and then brought in contact with oxygen. Nanoparticles form upon rapid quenching. The supersaturated vapor may be created using laser ablation, plasma torch synthesis, combustion flame, exploding wires, spark erosion, electron beam evaporation, sputtering, etc.

Temperatures as high as 10,000 K are typical of the resulting plasma during the laser ablation process. A high energy pulsed laser is shined on the target material and the material vaporizes rapidly. The process is expensive because of inefficiency of the laser.

Combustion flame synthesis and the plasma torch approach are used in large scale in the industry to manufacture nanoceramic powders. An oxidation atmosphere is used in the combustion flame synthesis method and the inert atmosphere is used in the plasma torch approach. In either of the processes, a raw material such as titanium tetrachloride ($TiCl_4$) is expensive. Titanium dioxide (TiO_2) is produced as a result of the combustion flame synthesis method, and titanium carbide (TiC) results as a product of the plasma torch approach. One of the problems that remain during manufacture of these products is the agglomeration of the ceramic nanopowders. The reactant feed materials are injected into a plasma jet in the plasma torch approach.

An aliter to this method of feed is by arc vaporization of the anode made of metal-ceramic composite. Less agglomeration can be seen in the plasma torch approach modified to effect rapid thermal decomposition of feed, use of a plug flow reactor (PFR), or rapid condensation of the product on a substrate that is cooled. Still the cost of production using the plasma torch approach is high. A quenching step is needed for the suspended particles as well as the gas medium in which the particles are suspended. An exploding wire device is used to synthesize nanoceramics. A wire with a diameter of 100 μm is Joule heated using a pulsed current that passes through it. The wire is vaporized and ionized and the resulting plasma undergoes rapid expansion. The contact surface of the plasma undergoes a reaction with the gas. This is followed by isentropic cooling. The mixing process depends on molecular diffusion. Complete vaporization requires smaller diameter wire and as a result, the rate of production is low.

The exploding wire method can also be conducted in a conventional capillary plasma device. The capillary plasma device consists of two noneroding electrodes positioned at the ends of a nonconducting bore with an open end. A fused wire connects the two electrodes. Explosive vaporization and ionization of the wire between the two electrodes is effected by an electrical discharge. The erosion and subsequent ionization of the liner maintains the discharge. Dense plasma inside the bore is produced. Then it exits from the open end of the device. Approximately 10,000 psi pressure has to be withstood by the sealed end or breech of the gun. The plasma that exits from the gun is used in the ignition of the explosive. The bore wall erosion is promoted by a large L/D ratio. This helps to sustain the discharge. The electrothermal gun is a pulsed power device for the production of high velocity plasma jets and vapors of different metals such as aluminum or titanium. A pulsed high current arc produces rapid vaporization of the electrodes and is struck down between an electrode in the breech of the gun and an electrode at the muzzle. Electrothermal gun processes are similar to the plasma torch approach. Electroguns can be used to atomize an external stream of molten metal. Electroguns have been used to synthesize nanomaterials with approximately 100 nm in diameter.

In order to provide an improved process, a hybrid exploding wire (HEW) method was developed by Nanotechnologies Inc., Austin, TX.[9] A nanoscale diameter solid metal fuse wire or foil sheath is used in the HEW process. The ends of the fuse are connected to electrodes that are designed for high erosion to sustain the heavy metal plasma. Large pressures are generated using the cylindrical construct when plasma is generated. The nanocrystalline ceramic powder is retained inside the bore. The process is started using an electrical discharge that triggers the vaporization and ionization of the fuse. The radial expansion of plasma is confined and upon exit from the ends of the tube, it reacts and mixes with a suitable gas in order to form nanoparticles. The bore wall is vaporized and plasma ablated, resulting in ceramic synthesis. Both physical and chemical processes are used to synthesize the nanoceramic particles in crystalline state. The problems typical of high-pressure breech seals are solved by exit of plasma on either end of the bore. The corresponding forces are removed due to internal pressures that are in place to balance each other. Increased production can be seen by concurrent operation of two electroguns. The same process can be conducted with the fused wire replaced with a conductive sheath.

The breech electrode is completely consumed. This is achieved by modification of the design of electrogun. A central electrode replaces the breech electrode. Two opposing electrothermal jets are produced that exit from either end of the electrogun. Two arcs are used to transfer electric current to the central electrode. There is no pressure drop between the two ends of the central electrode obviating the need for a seal. When a pressure drop develops between the two ends of the central electrode, there are no adverse consequences to the plasma leak and molten metal across the electrode.

It has been observed that nanoceramics exhibit superplastic properties. They deform in a ductile manner with remarkable elongations to failure such as up to 800 percent under tension at moderate temperatures. They can be used in near-net shape forming operations. Faster formation rates at lower temperatures are possible with the miniaturization of grain size. For example, nanocrystalline titania deforms super plastically at temperatures 300°C lower than achieved previously. Nanocrystalline yttria-stabilized zirconia exhibits superplastic strain rates 34 times faster than submicron grained yttria-stabilized zirconia. Application of nanoceramic materials to commercial forming operations have three key technical hurdles to be crossed: (1) large quantities of material need to be made available, (2) stock pieces such as plate, rod, bar, etc. used for superplastic forming operations should be densified, and (3) grain growth both during initial fabrication of stock pieces and later in the forming operations needs to be minimized. Super plastically formed jumbo jet components may need nanoceramic powder in ton quantities. Two processes to prepare such nanoceramics were discussed by Mayo, Hague, and Chen.[10] These processes are discussed in the following text.

Precipitation of Hydroxides from Salts

Metal hydroxides can be formed from water-soluble salts. Sparingly soluble metal hydroxides can be precipitated out into a nanoscale powder. Once the supersaturation of solubility limit is reached, a minor swing in temperature or concentration of the system will trigger the onset of precipitation. The precipitation consists of nucleation, embryo formation, growth, and formation of nanoparticles. The ionic concentrations and pH can be changed to decrease the final particle size. The hydroxide method is a more suitable chemical synthesis method for further processing into nanoceramics. Hydroxides easily decompose to oxides. The calcinations step can be conducted well below the sintering temperature of the ceramic. The interparticle diffusion or bonding step is less prevalent and agglomerates are not formed. Oxide ceramics can also be prepared from carbonates that form precipitates. However, the calcinations temperature is higher and the nanograin shape cannot be retained. For example, zirconium dioxide (ZrO_2) based ceramics can be crystallized by forming a solution of zirconium tetrachloride ($ZrCl_4$) or zirconium oxy dichloride

($ZrOCl_2$) and then later stirred with an ammonium hydroxide solution. Zirconium tetrahydroxide [$Zr(OH)_4$] precipitate is formed and complexes with water. The formed hydroxide is filtered, washed, and dried overnight in an oven at 110°C. This is calcined into ZrO_2 nanoparticles. The final particle is weakly agglomerated. The interparticle bonds are primarily van der Waals in nature. ZrO_2-Y_2O_3 powder with 3 mol% yitric oxide can be generated by coprecipitation of the $Zr(OH)_4$ with yitrium tetrahydroxide. Homogeneous nanoceramic particles can be seen under a TEM.

Sol-gel processes have been developed to produce the oxides directly from hydrolysis of metal alkoxides. During the first step, one or more alkoxy groups are removed and replaced with hydroxyl groups. Then a chain or network structure is developed when the hydroxyl groups act as attachment sites for other alkoxide molecules.

This process of monomer attachment via hydroxyl elimination is referred to as condensation. The rate-determining step can vary between the hydrolysis and condensation steps.

Pressureless Sintering

The nanocrystalline powders are first compacted and then they are sintered. Pressureless sintering involves heating the sample in an ambient atmosphere. Densification and grain growth processes occur at elevated temperatures and contribute to the production of crystalline nanoceramic particles for super plasticity. Both processes are diffusion-controlled phenomena and both occur concurrently.

Densification occurs in stages (Fig. 5.6). In Stage I, the interparticle necks form and in Stage II the ceramic can be modeled as a solid

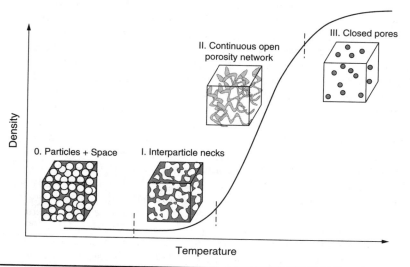

FIGURE 5.6 Three stages of densification of superplastic nanoceramic.

body containing an extensive interconnected network of continuous pore channels open to the outside surface. In Stage III, the open pores are eliminated leading to the formation of closed spherical voids.

5.9   Thermal Barrier Coating

Siemens Power Generation patented a ceramic thermal barrier coating material with nanoscale features with improved high-temperature performance.[11] Ceramic thermal barrier coatings tend to protect metal or ceramic matrix composite materials from oxidative corrosion in tough environments. Nickel- or cobalt-based super alloy components include a bond coat and a top coat of yttria-stabilized zirconia. This coating is applied using electron beam physical vapor deposition (EB-PVD) in order to constitute a column structure.

Magnesium oxychloride sorrel cement coatings can be used to delay the flame spread in naval warheads. The time taken for the transient temperature of the naval warhead to reach alarm temperatures was modeled by Sharma.[12] Thermal barrier coatings (TBCs) are advanced material systems applied to metallic surfaces operated at elevated temperatures. These coatings serve as insulators of heat. TBCs can be used in naval warheads. In the event of fire, metallic bulkheads and decks spread fire by heat transfer, which increases the temperature on the back side above the ignition temperature of common combustibles in adjacent and overhead compartments. In order to keep temperatures on the back side below the ignition temperature of the common combustibles, the navy uses fire insulation such as mineral wool or Structogard. The role of the TBC is to prevent fire spread. Coated onto a combustible substrate they can prevent the ignition and spread of flame. These can be used in applications with high fire potential. Magnesium oxychloride sorrel cement coatings have been developed by Delphi Research to be used as a barrier system. The ceramic coating can be prepared using the CVD or spray coating process.[13] The thermal conductivity of the coating may be reduced by increasing the porosity of the coat. Magnesium oxychloride cement coating cobonded with high alumina calcium aluminate cement colloidal silica was found to be effective. One of the critical parameters in the design of such systems is the resistance offered by the material to temperature rise fires. ASTM standards discuss this parameter in the UL 1709 test. The UL 1709 test lays down the performance criteria for insulation material. The back side average temperature rise should not exceed 250°F above the ambient when tested under the conditions prescribed in the test. UL-1709 fire is specified at 2000°F and 200 kw/m^2 for 30 min or more. The thrust of the research in development of TBC is looking to advance fire insulation technologies and bring in new materials that reduce weight and cost.

The nature of heat transfer during a fire spread is transient. The insulation materials are usually heterogeneous in nature. Under short timescales, Fourier's law of heat conduction does not accurately depict the real events.[14] Mathematical models that can predict the time of rise of insulation temperature specified in UL 1709 that use Fourier's parabolic law of heat conduction cannot be relied upon. This is because of the presence of "blow-up" or singularities in the expression for heat flux in the analytical solution obtained using Fourier's law. Damped wave conduction and relaxation may become significant during the short time scales expected when TBC is used to delay onset of fire. It cannot be neglected. The analytical solution for temperature profiles in the naval warhead and in its thermal barrier coating layer, TBC, is derived for the naval warhead layer and the coating layer by the method of separation of variables. The external fire temperature, T_{fire}, drives the problem. The final condition in time is used as the fourth constraint and the second time constraint to obtain bounded and meaningful solutions. These solutions do not disobey Clausius' inequality. The solutions were obtained by the method of separation of variables. A critical coating thickness was derived, below which the transient temperature in the coating will undergo subcritical damped oscillations. This will drive oscillations in temperature within the naval warhead. In the literature, coatings are usually by design made as "thin" as possible. The results of this study clearly show that they cannot be made too thin. A desirable mode of operation is when the transient temperature does not oscillate and hence the coating thickness has to be greater than the critical thickness of coating as predicted. In a similar manner, the naval warhead cannot be made of materials with relaxation time greater than $R^2/15.33\alpha_{nr}$ or else the temperature in the naval warhead will undergo subcritical damped oscillations in time.[15] The exact solutions for transient temperatures within the naval warhead and in the TBC are seen to be bifurcated. Both solutions are given by an infinite Bessel series of the first kind and zeroth order and decaying exponential in time. Under some conditions, the time domain portion of the solution becomes cosinous.

5.10   Ceramic Nanocomposites

Roy, Komarneni et al.[16] coined the term *nanocomposite*. They prepared hybrid ceramic metal nanocomposite materials synthesized by sol-gel processes. Their quasi-crystalline materials had phases of 1 to 10 nm regions with different structure. Materials with more than one Gibbsian solid phase where at least one of the dimensions is in the nanoscale region are called nanocomposite materials. The state of the material can be amorphous, crystalline, or semicrystalline. Different categories of ceramic nanocomposites include the following:

1. Structural ceramic nanocomposites such as MgO/SiC, mullite/ SiC, Al_2O_3/SiC, and MgO/SiC

2. Glass ceramics such as photosensitive glasses

3. Glass/metal nanocomposites

4. Electroceramic nanocomposites such as Co/Cr

5. Film nanocomposites such as lead zirconate titanate/nickel

6. Entrapment-type nanocomposites such as zeolite/metallic or zeolite/organic complexes

7. Layered nanocomposites such as pillared clays

8. Metal/ceramic nanocomposites such as Fe-Cr/Al_2O_3

9. Organoceramic nanocomposites such as polymeric matrix/ $PbTiO_3$

10. Low temperature sol-gel derived nanocomposites such as Al_2O_3/SiO_2, AIN/BN, and MgO/SiC

Ceramics are brittle. The fracture strength of brittle materials can be improved by an increase in fracture toughness or by critical reduction in flaw size. The size and density of processing flaws are decreased using improved processing technology. Energy dissipating components such as whiskers, platelets, or particles can be included in the ceramic microstructure to increase the fracture toughness of the material. The crack is deflected by the included components. Metallic ligaments may also be incorporated into the matrix. There is considerable work undertaken in the literature in optimization of structural nanocomposites with emphasis on matrices made out of alumina (Al_2O_3) with nanoscale silicon carbide (SiC) particles dispersed in them. The nanosized SiC is introduced as second phase. The interesting properties of nanocomposites are wear, creep, and high temperature performance.

Sol-gel processing, polymer processing, and traditional powder processing can be used to prepare Al_2O_3/SiC nanocomposites.[17] The traditional powder processing method can be divided into four main steps: (1) selection of raw materials, (2) wet mixing of powders, (3) drying the slurries, and (4) consolidation. γ-Al_2O_3 and α-Al_2O_3 powders are used as raw materials for the matrix. The raw material powders must possess small particle size and have high purity levels. Powders with less purity will result in the formation of second phase during sintering. The alumina powders have loosely packed morphology and possess high surface area. Wet ball milling is used to achieve homogenization of the powder mixture. Slurries with water and organic media differed in their powder stabilization. Methanol promotes deflocculation. Modified powders can be used to improve the properties of the slurries. Ultrasonic dispersion and sedimentation have been used to break up the agglomerates of SiC powder. Drying of the slurries is a critical step and minimizes the risk of formation of agglomerates. One team of investigators used an infrared heat lamp for the drying step. Hard agglomerates are formed when the drying rates are high at high temperatures. Freeze-drying has been found to be an optimal method for removing moisture from aqueous suspensions. Agglomerate formation is suppressed in the short time necessary for freezing and segregation is avoided. Dried powders have been calcined at 600°C for 10 h in air.

Hot pressing is used for consolidation of nanocomposites. It is carried out using graphite die, with pressures of 20 to 40 MPa and temperatures between 1550 and 1800°C under argon or nitrogen atmospheres. The temperatures needed increase with increase in vol% SiC: 1600°C for 5 vol% SiC; 1700°C for 10 vol% SiC; 1800°C for 17 vol% SiC. Slip casting, injection molding, and pressure filtration methods can be used for densification of up to 62 percent. TEM micrographs of Al_2O_3/SiC nanocomposites reveal a pinning effect and not so straight grain boundaries. Occurrence of dislocations can also be seen. Dislocations are generated using cooling down from the fabrication temperature. This may be due to the high internal stresses near SiC particles. Post annealing can result in further increase of the strength of the material.

β'-sialon can form from impurity catalyzed reactions. Decomposition reactions can occur in the nanocomposite during sintering. These are some of the thermal stability issues of nanocomposites. Nitrogen molecules can dissociate and chemisorb on the surface of the SiC particles. This reaction may be catalyzed by the metallic impurities and form Si_3N_4.

$$3SiO_2 + 2N_2 \rightarrow Si_3N_4 + 3O_2$$

Mullite may also form at high temperatures by oxidation of the SiC particles in the surface layer followed by reaction with alumina. Thermodynamic studies of the Al_2O_3/SiC nanocomposite system reveal that at temperatures above 1700°C the powders are not stable. Heavy weight losses can be observed during sintering because of volatilization and decomposition of SiC. A gaseous mixture of CO, SiO, and Al_2O can form when Al_2O_3 and SiC react with each other. At temperatures above 1950°C a liquid can form from reaction of SiC with Al_2O_3. The liquid phase consists of aluminum (Al). Otherwise, a eutectic mixture of Al_2O_3-Al_4C_3 or SiC-Al_4C_3 eutectic forms.

A second method of preparing the alumina/silicon carbide nanocomposites is by pyrolysis of silicon containing polymeric precursor. The agglomeration and dispersion problems experienced in the classical powder processing method can be improved upon using this method. Polycarbosilane is coated onto a surface-modified alumina powder and pyrolyzed at 1500°C to produce ultrafine SiC particles with particle size less than 20 nm. A densified nanocomposite is formed by hot pressing the powder at 1700°C. A fine dispersion of SiC is effected.

The sol-gel processing method can be used to prepare alumina/silicon carbide nanocomposites by use of behomite gels as the source of alumina. The gels are coated on silicon carbide particles. The behomite powder is deagglomerated and mixed with distilled water to form a transparent sol. Drying, calcinations, and hot pressing are the subsequent steps carried out. Smaller grain sizes are achievable using the sol-gel processing method. An at-a-glance side-by-side comparison of the conventional powder processing method, polymer coating method, and sol-gel method is given in Fig. 5.7.

Nanocomposite materials prepared from hot pressing can be seen to have improved mechanical properties. The strength was found to increase with relative density and decrease with increase in porosity. Unwanted secondary reactions can lead to deterioration of properties. Hardness of the material was found to follow a linear rule of mixture as a function of the silicon carbide content. Hardness of 1.75 GPA and Young's modulus of 404 MPa can be observed. A maximum was

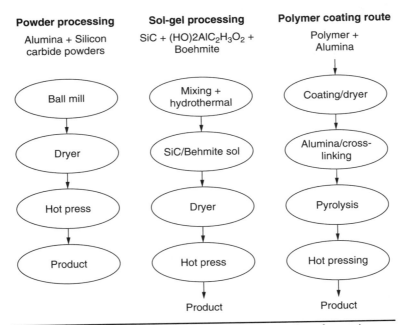

FIGURE 5.7 Flow chart for the three methods for preparation of ceramic nanocomposites.

observed in the strength and toughness as a function of SiC content of the product. The *fracture strength* of the nanocomposite materials was found to vary from 350 to 560 MPa. Addition of 5 vol% SiC nanoparticles can lead to increases in strength of up to 1050 MPa. Further increases in dispersant can lead to a decrease in the fracture strength down to 800 MPa. Sometimes the decrease in strength is due to agglomeration problems. The matrix grain size of the product decreases with decreases in sintering temperature but also increases with increases in SiC content. The Hall-Petch equation can be used to predict the strength of ceramics. The *bend strength* of the nanocomposite can be improved by a factor of 50 percent by annealing the material for 2 h at 1300°C under argon atmosphere. This may be due to the healing of machine-introduced cracks at the tensile surface of the bend test beams or due to the relaxation of internal stresses. The *fracture toughness* of the nanocomposites was found to increase dramatically at 5 vol% SiC. The property tends to decrease or plateau upon further increases in vol% SiC. Fracture toughness is measured using indentation precrack followed by bend testing. Fracture toughness of nanocomposites was found to be approximately 100 percent higher than the toughness of alumina when indentation crack length measurements were used.

The use of SiC nanoparticles as dispersants results in increase in wear resistance, creep resistance, and high temperature strength. Wear rates can be measured using erosion measurements. The matrix grain size has an effect on wear rate. The grain boundary fracture is inhibited during wear. Fine-grained alumina undergoes plastic deformation resulting in higher wear rates. The most important creep resistance mechanism is suspected to be grain boundary sliding and dislocation movement. SiC nanoparticles inhibit these.

5.11   Summary

The high impact of nanostructuring operations in society has lead to undergraduate institutions adapting these advances into their curricula in order to better train the engineering students who are expected to work in the nanotechnology industry. Optical transform slides have been used to illustrate x-ray diffraction. Students measure the feature spacing on a slide using a plastic ruler and hand lens, and compare their results with those obtained in a simple diffraction experiment. Ferrofluids are prepared in the nanostructuring laboratory. Surfactant is used to control the degree of agglomeration of particles of magnetite. Principles of oxidation state, stochiometry, crystal structure, magnetism, and surfactant chemistry are illustrated using this kind of experiment. Nickel nanowires that are 200 nm in diameter and 50 μm in length can be prepared using a template synthesis technique. They are analyzed using SEM and powder x-ray diffraction techniques. Redox chemistry, colloids, electromagnetic radiation, and polarization effects are illustrated using a colloidal gold experiment. Sodium citrate is reacted with hydrogen tetrachloroaurate. Microfluidic technology, SAM deposition, and AFM, can be demonstrated using an experiment where nanoparticles are filtered from a solution.

Polymer nanocomposite is emerging as a new vista in materials science. Examples are development of exfoliated clay nanocomposites, CNTs, exfoliated graphite, nanocrystalline metals, and other filler modified composite materials. Performance enhancement includes increased barrier properties, flammability resistance, electronic properties, membrane preparation and polymer blend compatibilization, and increased coefficient of thermal expansion. Clay platelets are 9.4 A^0 in thickness. Nanocomposite can exhibit immiscible, intercalated, and exfoliated morphologies. WAXS spectra are used to relate morphological structure to the performance properties of nanocomposites.

Aqueous magnetic ferrofluid compositions are used in magnetic imaging and printing with dry and liquid developer compositions, electrophotography, xerographic imaging, and printing. It is a colloid and polarizes in the presence of a magnetic field. Fe_2O_3 nanoparticles are milled down to less than 10 nm and an aqueous liquid is added to form a colloidal suspension.

SMAs are metal-based alloys. Metals used are copper, iron, and nickel. Solid-state transformation called martensitic transformation is associated with SMA. These materials exhibit shape memory effect and exhibit a superplastic state. Transition from the martensite phase to the austensite phase is dependent only on temperature and stress, and not on time. Reading glass frames are made of SMA.

Nanowires possess diameters in the range of 1 to 500 nm. Photoemission of semiconductor nanowires, magnetic behavior of Ni and Co nanowires, current-voltage response of platinum nanowires are examples of application of nanowires. Nickel nanowires have been patterned on SiO_2 substrate. Nanowires have superior thermoelectric properties.

Liquid crystals are ordered at the nanoscale but lack long-range order. Cholesteric phase liquid crystals contain molecules that are aligned in layers rotated with respect to each other. As the temperature of a liquid crystal is increased, all the colors of the rainbow can be observed from red to orange to yellow to green to blue. Maghemite (γ-Fe_2O_3) nanoparticles can be used as a superhigh density magnetic recording substance due to their shape anisotropy and magnetic characteristics. Maghemite nanoparticles with rod-shaped morphology with a diameter of 250 nm and an L/D ratio of six are used in high-density magnetic recording substances. Nanoparticles with less than 10 nm exhibit super magnetism.

Nanoceramics possess excellent heat, corrosion, and abrasion resistance and exhibit unique electrical, optical, and magnetic properties.

Nanocrystalline ceramics with 1 to 500 nm diameters have been reported in the literature. The three different methods used in the synthesis of nanocrystalline ceramics can be classified into chemical processing, thermal processing, and mechanical processing. Nanoceramics exhibited superplastic properties. Ceramic thermal barrier coatings with nanoscale features with improved high temperature performance have been developed. Magnesium oxychloride sorrel cement coatings can be used to delay the flame spread in naval warheads.

Ceramic metal hybrid nanocomposites synthesized by sol-gel process were found to have phases of 1 to 10 nm quasi-crystalline regions with different structures. Al_2O_3/SiC nanocomposites can be prepared by sol-gel processing, polymer processing, and traditional powder processing.

Review Questions

1. What parameters does the Fraunhofer equation relate?

2. What is the connection between x-ray diffraction and optical transform slides?

3. What is the role of surfactant in preparation of ferrofluids?

4. What is the role of an AA battery during the preparation of nickel nanowires?

5. What is meant by the Tyndall effect?

6. What reduction reaction is responsible for a reddish colloidal suspension of gold nanoparticles?

7. Identify an interfacial reaction that causes the formation of a nanofilter.

8. What is the significance of phase separated polymer blends in creating nanostructures?

9. Nanoscale void structures are present in what membrane?

10. What are the conditions favoring spherulite growth?

11 What are the characteristic features of exfoliated clay nanocomposites?

12. What is meant by confined crystallization?

13. Show by using a neat schematic the different layers with dimensions that constitute montmorillonite clay intercalated by polyolefins.

14. What are the differences between exfoliated and intercalated nanocomposites?

15. What are the differences in morphology of immiscible, intercalated, and exfoliated nanocomposites?

16.   How is the WAXS spectra used to distinguish between exfoliated and intercalated nanocomposites?

17.   Addition of fillers to polymers results in what in the modulus or stiffness by reinforcement mechanisms?

18.   What happens to HDT when montmorillonite is added to a nylon 6 matrix?

19.   Comment on the coefficient of thermal expansion property of nanocomposites.

20.   What happens to the barrier properties of exfoliated clay modified PET?

21.   How is the permselectivity of membranes improved by clever nanocomposite design?

22.   How are nanopores generated in a scaffold structure that is bioresorbable?

23.   What nanocomposites were introduced by Toyota as belt covers?

24.   What nanocomposites were developed by General Motors for use in exterior trim of vans?

25.   What nanocomposites are used in (a) tennis rackets, (b) hockey sticks, (c) tennis balls, and (d) electrostatic dissipation?

26.   What nanocomposites were introduced as (a) bandages, (b) beverage containers, (c) winter tires, and (d) auto fuel systems?

27.   What happens to ferrofluids in the presence of a magnetic field?

28.   Distinguish ferrofluids from magnetorheological fluids.

29.   Do ferrofluids possess a spike forming property? Why or why not?

30.   Discuss the applications of ferrofluids such as in a nanocompass.

31.   What is meant by superplasticity?

32.   What is the role of martensitic transformation in superplastic behavior?

33.   Discuss the shape memory effects in materials and the role of nanostructuring in preparation of SMAs.

34.   How are SMAs used in reading glass frames?

35.   What are the manufacturing methods for preparation of SMA materials?

36.   How are SMA materials used in (a) adhesion of dissimilar metals, (b) coupling of oil pipelines, (c) vascular stents, (d) dental braces, (e) nitinol wires, and (f) robotics?

37.   At what diameter do nanowires expect to display quantum conduction phenomena?

38. What are the different methods for preparation of nanowires?

39. What is the role of pyrolysis in the Honda process for preparation of nanowires?

40. At what dimensions do nanowires display semiconductor properties?

41. Where are quantum confinement effects found?

42. What is the significance of observation of ballistic transport?

43. Show a qualitative phase diagram of bismuth nanowires.

44. Discuss the stacks of layers in cholesteric liquid crystal and their role in observation of the colors of a rainbow.

45. What is the niche property of γ-Fe_2O_3 nanoparticles?

46. At what nanoparticle size does super magnetism set in?

47. What is the advantage of changing inherent paramagnetic spherical FePt particles into ferromagnetic particles?

48. Why do rod-shaped particles exhibit ferromagnetism?

49. What is the difficulty encountered when attempting to synthesize nanoparticles with less than 30 nm diameter?

50. How are reverse micelles used in the generation of nanoparticles?

51. What are the three broad classifications of the synthesis methods for nanocrystalline ceramics?

52. Compare the arc vaporization method and combustion flame synthesis method used for preparation of nanoceramics.

53. In which methods are rapid quenching of superheated vapor used?

54. What are sol-gel techniques?

55. What is the role of an electrothermal gun in the exploding wire method of synthesis of nanoceramics?

56. What are the process improvements effected by the HEW method?

57. What are the three key technical hurdles to applying nanoceramic materials to commercial forming operations?

58. Discuss the process of precipitation of hydroxide from salt to form nanoceramic powder used in super plastically formed jumbo jets.

59. How is densification affected in the pressureless sintering method to form nanoceramic powder in super plastically formed jumbo jets?

60. Discuss the design of a ceramic thermal barrier coating in terms of its chemical composition, application, performance enhancement, and microstructure.

61. How are magnesium oxychloride sorrel cement coatings used in flame retardant applications?

62. What are the different categories of ceramic nanocomposites?

63. How is the fracture toughness of ceramics improved by reduction in flaw size?

64. Discuss the optimization of structural nanocomposites made of ceramic materials.

65. Distinguish sol-gel processing methods to prepare alumina/SiC nanocomposites with the pyrolysis method.

66. Discuss the chemical stability at high temperature of SiC ceramic systems.

67. Discuss deterioration of properties due to unwanted secondary reactions.

68. How are improvements affected in performance of nanoceramics in the following categories: (a) hardness, (b) fracture strength, (c) bend strength, and (d) fracture toughness?

69. Name two applications of nanostructuring that have a strong impact on society.

References

1. A. K. Bentley, W. C. Crone, A. B. Ellis, A. C. Payne, K. W. Lux, R. W. Carpick, D. Stone, G. C. Lisensky, and S. M. Condren, Incorporating concepts of nanotechnology into the materials science and engineering classroom and laboratory, *Proc. 2003 ASEE Annu.Conf. Exposition,* 1462, 2003.
2. Y. L. Loo, R. A. Register, and A. J. Ryan, Polymer crystallization in 25 nm spheres, *Phys. Rev. Letters.,* 84, 4120–4123, 2000.
3. D. R. Paul and L. M. Robeson, Polymer nanotechnology: Nanocomposites, *Polymer,* 49, 3187–3204, 2008.
4. R. F. Ziolo, E. C. Kroll, and R. Pieczynski, High Magnetization Aqueous Ferrofluids and Processes for Preparation and Use Thereof, US Patent 5,667,716, 1997, Xerox Corp., Stamford, CT.
5. A. Harutyunyan, Methods for Synthesis of Metal Nanowires, US Patent 7,341,944, 2008, Honda Motor Co., Minato-Ku, Tokyo, Japan.
6. M. S. Dresselhau, Y. M. Lin, O. Rabin, A. Jorio, A. G. Souza Filho, M. A. Pimenta, R. Saito, G. G. Samosonidze, and G. Dresselhaus, Nanowires and nanotubes, *Mater. Sci. Eng.,* C23, 129–140, 2003.
7. G. C. Lisensky, D. Horoszewski, K. L. Gentry, G. M. Zenner, and W. C. Crone, Fats, oils, and colors of a nanoscale material, *The Science Teacher,* 30–35, December 2006.
8. K. Woo, J. P. Ahn, and H. E. Lee, Shape Anisotropic Metal Oxide Nanoparticles and Synthetic Method Thereof, US Patent 7,122,168, (2006), Korea Institute of Science and Technology, Seoul, Korea.
9. D. R. Peterson, D. E. Wilson, and D. L. Willauer, Method and Apparatus for Direct Electrothermal-Physical Conversion of Ceramic into Nanopowder, US Patent 6,600,127, (2003), Nanotechnologies, Inc., Austin, TX.
10. M. J. Mayo, D. C. Hague, and D. J. Chen, Processing nanocrystalline ceramics for applications in super plasticity, *Mater. Sci. Eng.,* A166, 145–149, 1993.

11. A. J. Burns and R. Subramanian, Thermal Barrier Coating having Nanoscale Features, US Patent 7,413,798, 2008, Siemens Power Generation, Inc., Orlando, FL.

12. K. R. Sharma, On modeling stability of magnesium oxychloride sorrel cement thermal barrier coating, AIChE Spring National Meeting, New Orleans, LA, 2008.

13. K. R. Sharma, Critical thickness of coating and onset of sub critical oscillations, 231st ACS National Meeting, Atlanta, GA, 2006.

14. K. R. Sharma, Time taken to alarm in high temperature barrier coating, AIChE Spring National Meeting, New Orleans, LA, 2003.

15. K. R. Sharma, Some issues in thermal barrier coating, CHEMCON 2004, Mumbai, India, 2004.

16. S. Komarneni, Nanocomposites, *J. Mater. Chem.*, 2, 1219–1230, 1992.

17. M. Sternitzke, Review: Structural ceramic nanocomposites, *J. Eur. Cer. Soc.*, 17, 1061–1082, 1997.

CHAPTER 6

Nanotechnology in Life Sciences

Learning Objectives

- Introduce concepts of molecular computing
- Discuss molecular machines and molecules that show promise
- Concepts of supramolecular chemistry
- Nanochips and gene sequencing measurements
- Sequence alignment, dynamic programming, string algorithms
- HMMs and applications, gene finding methods
- Protein secondary structure prediction
- Polymer nanoencapsulation and drug delivery systems

6.1 Molecular Computing

When the limits of miniaturization of transistor packing and gate width in silicon chips are reached, where will further increases in microprocessor speed come from? They can come from *biochemical nanocomputers*. The field of DNA computing originated when a programmable molecule-computing machine composed of enzymes and DNA molecules was unveiled in 2003 at the Weizmann Institute of Science in Rehovot, Israel. The computer operations were at a rate of 330 terraflops. This was 100,000 times faster than a personal computer. It was entered into the Guinness Book of World records as the smallest biological computing device ever constructed. Molecular computing is expected to emerge when the limits of miniaturization in silicon chips are realized as the key to further increases in computing speed. In 1994, the idea to use DNA molecules to store and process information took shape when a scientist from California used DNA in a test tube to solve a simple mathematical problem. Designs of DNA computers were drawn up where adenosine triphospate (ATP) molecules where thrown in to provide a steady supply of energy and fuel. The enzymes served as the hardware and the DNA served as the software. The ways in which molecules undergo chemical reactions with each other allow simple computer operations to be performed as a by-product of the reactions. Scientists program the devices by controlling the composition of DNA software. One trillion biomolecular devices can be fit into a single drop of water. Results are analyzed by a method where the length of the DNA output molecule is seen in place of a computer monitor. This computer performs rudimentary operations. Self-organization of molecules is used in the design of molecular computation. This gave rise to the field of *molecular electronics*. Processing in biological systems is used to devise nanomachines. The autonomous formation of complex nanostructures is considered one type of computation. In a cellular computer, membrane proteins are expected as input/output (I/O) devices. Living cells are tapped into and the signal transduction functions are used for computations.

DNA computing was born in 1994 when Adleman[1] proposed a molecular algorithm to solve the Hamiltonian path problem with DNA and solved an instance of a directed graph with seven nodes. The parallelism of DNA molecules was exploited. This offered a faster solution of nondeterministic polynomial (NP) complete problems. Reactions at the nanoscale are used to perform computations with less energy. Several nanofabrication techniques were developed in the area of DNA nanotechnology. Scientists are beginning to obtain a better handle on the electric charge distribution in DNA and the charge transfer observed in DNA is used in a device of novel molecular electronic circuits. The automata and the computational model are implemented using hairpin-formed DNA molecules. The polymerase chain reaction (PCR) that is found during DNA transcription and translation is tapped into. Autonomous assembly of molecular structures is used in devising DNA computation in the solid phase. The number of DNA molecules needed for effecting automated molecular computation is calculated. Yoshida and Suyama[2] studied the application of a DNA computer to biotechnology such as gene expression analysis and molecular memory. In *aqueous computing*, the write once memory is represented by double-stranded circular DNA. Plasmid contains multiple regions whose terminals are flanked by restriction sites. The write operation is implemented by removing a particular region using a specific restriction enzyme. Head, Yamamura, and Gal[3] proposed the molecular solution with write once memory to obtain solutions of NP complete problems as Max-Clique.

6.2   Molecular Machines

Molecular machines are molecules that, with an appropriate stimulus, can be temporarily lifted out of equilibrium and can return to equilibrium in the observable macroscopic properties of the system. There is considerable debate in the literature as to the exact constitution and properties of molecular machines. The controlled motions of synthetic molecular systems have been harnessed to cause observable macroscopic changes in bulk systems because of *stereochemical* rearrangement at the molecular level. The following have enabled construction of molecular machines:

1. Progress in organic synthesis—living free radical polymerizations, asymmetric catalysis, metal catalyzed cross-coupling reactions, and metathesis
2. Powerful computational techniques
3. Advent of collection of powerful single molecule analytical tools

Molecular components are constructed in such a fashion so they interact favorably with each other and so they can *self-organize* and *self-assemble* into larger well-defined architectures. The advances in spectroscopic techniques are also tapped into.

The Nobel Prize in 1998 for chemistry went to Pople and Kohn for computing accurately many physical and electronic properties that have particular relevance to molecular machines and electronics. SAMs, Langmuir-Blodgett techniques for creation of monolayers, soft-lithographic techniques, etching the surface, AFM, scanning tunneling microscopy (STM), x-ray photoelectron spectroscopy (XPS), photoelectron spectroscopy, and ellipsometry and x-ray reflectometry are used in the manufacture of molecular machines.

It is difficult to create molecular actuators. The Feynman ratchet can be a violation of the second law of thermodynamics and may be a PMM1 or PMM2 perpetual motion machine. Actuation can be achieved by a walking mechanism using a class of motor proteins called *kinesins*. The muscle contraction and expansion involves such chemical changes. ATP hydrolysis and whip-like and sinusoidal movements of cilia and flagella enable them to transport cells throughout the body. Based on these observations, molecular machines, molecular shuttles, switches, muscles, nanovalves, rotors, and surfaces with controlled wettability are being created. Chemicals involved in such constructs are, for example, rotaxane molecules. Calixarenes in the cone formation have an interval cavity able to host guest molecules of complementary size. The inclusion of guests in solid state has been studied in an apolar medium. The solvation phenomenon is of interest. The host-guest association is governed by coulombic attractions such as: (1) steric, (2) entropic, and (3) solvation. Rotaxane synthesis involves calixarene wheels. During the synthesis of pseudorotaxanes, calixarene derivatives are hosts for QUATS, a triphenylureidocalixarene derivative. The structure of pseudorotaxene can be studied using x-ray techniques.

A *molecular shuttle* consists of a ring component that is mechanically interlocked onto a dumbbell-shaped component and is able to shuttle between two recognition sites because of thermal activation by noncovalent interactions. *Molecular switches* are compounds that can be externally stimulated to exist in either of two observably different states or conformers. A *molecular muscle* is able to expand and contract reversibly upon external stimulation. *Molecular valves* can be used to trap and release other molecules because of controlled molecular motions. *Molecular rotors* undergo controlled rotational motion of a rotor until relative to a stator, which is controlled via an axle. The surfaces with controlled wettability can be stimulated to be hydrophobic or hydrophilic. Mechanical movements of cyclobispara-quat (p-phenylene), CBPQT along a diaminobenzene-containing thread that is tethered to a gold surface can be observed. A two-electron reduction of the CBPQT ring erases the favorable binding interactions that exist along the ring host and diaminobenzene guest with an electron transfer rate of 80 Hz. The oxidative electron transfer rate is 1100 Hz. Reversible redox-controllable mechanical motions of the interlocked molecule on a gold surface can be detected. A contracted structure can be generated by contraction of a bistable dimer by quantitative demetalation of the cuprous ions using potassium cyanide (KCN) followed by treatment with $Zn(NO_3)_2$.

Technologists are developing *molecular nanovalves*, irreversible thin film regulators, light regulated azobenzene and coumarin valves, supramolecular gatekeepers, reversible nanovalves, polymeric valves, biological nanovalves, etc. The three fundamental molecular motions are translation, rotation, and vibration. Scientists at University of California at Los Angeles (UCLA) have demonstrated controlled unidirectional rotation of a tryptycene rotor unit relative to a helacine stator that is connected by a carbon-carbon single bond axle using phosgene as a chemical fuel. The difluorophenylene rotor has been used as a molecular compass and a gyroscope.

6.3   Supramolecular Chemistry

Supramolecules are highly complex chemical systems made from components interacting through noncovalent intermolecular forces. The field of supramolecular chemistry occurs in the interface of biology and physics.[4] Strands of nucleic acids allow for huge amounts of information to be stored, retrieved, and processed via weak hydrogen bonds. The principles in molecular information in chemistry were developed from these observations. Interactional algorithms were developed through molecular recognition events based on well-defined interaction patterns such as hydrogen bonding arrays, sequences of donor and acceptor groups, and ion coordination sites. The goal here is to gain progressive control over complex spatio-structural and temporal dynamic features of matter through self-organization. The design and investigation of preorganized molecular receptors that are capable of binding specific substrates with high efficiency and selectivity are undertaken. The three themes are: (1) molecular recognition, (2) self-organization, and (3) adaptation and evolution.

Supramolecular materials offer alternatives to top-down minia-turization and bottom-up nanofabrication. Controlled assembly of ordered, fully integrated, and connected operational systems effects self-fabrication by hierarchical growth. The field of adaptive/evolutive chemistry emerged. Adaptive chemistry implies a selection and growth under time reversibility. The era of Darwinistic chemistry has dawned. The goal here is to merge design and selection in self-organization to perform self-design in which function driven selection among suitably instructed dynamic species generates the optimal organized and functional entity in a post-Darwinian process. Chemical learning systems cannot be instructed but can be trained. Time is irreversible. The passage is from closed systems to open and coupled systems that are connected spatially and temporally to their surroundings. Inves-tigators are progressively unraveling the complexification of matter through self-organization.

6.4   Biochips

Genetic studies can be conducted using microarray analytical devices in an unprecedented manner at a great speed and precision. mRNA labeled from cells, tissues, and other biological sources are examined using fluorescent probes in glass chips packed with thousands of genes. Spots on the biochips glow because of the reaction between the fluorescent probes and gene sequences. The stronger the activity of the gene expression, the greater is the intensity of the glow. The cost of gene expression has been reduced from several thousands of dollars to a few thousand dollars per gene scanned with advances in mea-surement technology. In one single experiment, the entire genome can be studied. Human disease, aging, drug and hormone action, mental illness, diet, and many other clinical matters can be studied using microarrays. This is because the patterns of gene expression correlate strongly with function. Alterations in gene sequences can be detected using microarrays. An era of genetic screening, testing, and diagnostics has dawned. Traditional histological and biochemical assays are obtained by miniaturization of tissue and protein microar-rays. Analysis of tumor specimens, protein-protein interactions, and enzymes are speeded up using microarrays.

Specific binding of genes and gene products on a substrate that is planar is allowed for in an ordered array of nanoscopic elements called a nanoarray. Earlier, due to the micron-sized features of the dots, they were called microarrays. The term *microarray* is coined from Greek and French words: *mickro* in Greek means small and *arayer* in French stands for arranged. Other words used for microarray are biochips, gene chips, DNA chips, protein chips, etc. Schena, who has done some pioneering work along with his professor R. Davis, has laid down the criteria for qualification as a microarray:

Ordered

Microscopic

Planar

Specific

When analytical elements are arranged in rows and columns, they qualify as ordered arrays. Uniformity of size and spacing of the ordered elements are important. The columns form a vertical line and the rows form a horizontal line along the slide. Any item with a majority of dimensions less than 1 mm is considered to be microscopic.

Photolithography is used to produce microarrays with micron-sized features. Spots of 200 to 600 μm can be seen using tissue microarrays. A DNA spot contains more than one billion molecules attached to a glass substrate. The elements of a microarray slide may typically consist of collections of target and probe molecules bound specifically to the substrate. These molecules may be gene or gene products. DNA, cDNA, mRNA, protein, small molecules, and tissues are examples of target materials used. These allow for quantitative analyses. They may also come from natural and synthetic derivatives. The sources may be cells, enzymatic reactions, and machines that carry out chemical synthesis. Shorter length nucleotides than the polynucleotide from a length of 2 bp to less than one billion are considered *oligonucleotides*.

Oligonucleotides make excellent target materials. The trend is to increase the density from 5000 elements per square centimeter to even greater packing efficiencies. Rapid kinetics are enabled and the entire genome is studied. Miniaturization and automation techniques are used. Filter arrays that do not permit miniaturization cannot be used for whole genome analysis.

Glass, plastic, and silicon can be used as planar substrates. This is where the microarray is configured. Glass has been used with success in preparation of microarray slides. Planar substrates offer several advantages compared with silica substrates. The materials are flat, spanning the whole surface. Automated manufacture is enabled by flat supports. Accurate scanning and imaging are possible and uniform detection distance between the substrate and detector is made possible. They are impermeable to liquids. They allow for smaller feature sizes and handle lower reaction volumes. Unique biochemical interactions between probe molecules in solution and the target molecules on the microarray allow for specific binding to the substrate. The most accurate measure of genes or gene products are provided when each microarray spot/target binds essentially to a single species in the labeled probe mixture. A target-one probe molecule paradigm is achieved using microarray assays. Assay precision can be enhanced using multiple microarray elements per gene. The minimal target length required to achieve single gene specificity is 15 to 25 bp nucleotide target sequence length.

The combined expertise from biology, chemistry, physics, engineering, mathematics, and computer science is tapped into by microarray technology. Pauling made the first correlation between gene mutation, altered protein, and disease. Hemoglobin from sickle cell patients was found to differ from hemoglobin in healthy individuals. This was detected by aberrant migration during studies of gel electrophoretic assays. This observation was attributed to a change in the surface charge of the molecule. Pauling concluded correctly after examination of normal individuals, carriers, and sickle cell patients that changes in the hemoglobin gene were responsible for the altered protein. This was verified later using gene sequencing studies. The dissemination of this work formed a landmark journal article that founded the field of molecular genetic analysis of human disease. It started the use of microarrays in genetic screening, testing, and diagnostics.

Microarray hybridization reactions are based upon the double helical structure of DNA discovered by Watson, Crick, and Williams. PCR reaction is used in microarray analysis. The enzyme DNA polymerase[5] exhibited catalytic activity. RNA polymerase and reverse transcriptase are tapped into in nanoarray studies. Baltimore[6] found that reverse transcriptase catalyzes the synthesis of DNA. Kornberg won the Nobel Prize recently for elucidating the molecular basis for transcription. DNA sequencing technology was developed by Maxam and Gilbert[7] and Sanger.[8]

There are some parallels between the emergence of the biochip industry and the silicon chip revolution. Schockley is called the father of Silicon Valley, was a cocreator of the transistor, and received the Nobel Prize for physics in 1956. He was the founder of Shockley Semiconductors in Palo Alto, California. A group of disgruntled employees called the Traitorous Eight resigned from Fairchild Semiconductor and floated a new company that manufactured integrated circuits. Noyce and Moore were among the eight. Silicon-based fabrication methods were selected and the first ICs were brought to the market in 1961. Moore's law came about in 1965. Moore was the head of research and development at Fairchild. He noted that transistor density and computing power doubles every 12 to 18 months. Noyce and Moore, who quit Fairchild Semiconductors, started Intel. The commercial Intel microprocessor came about in 1971. The Intel 4004 silicon chip contained a 2300 transistor. These were capable of performing 100,000 calculations per second at 108 KHz. Greater computing power can be seen in modern silicon chips. Forty-two million transistors were packed into a Pentium IC chip. This was released in 2000 with a capability of performing 1.5 billion calculations per second at 1.5 GHz. They were made with 180-nm circuit lines.

In a similar fashion, the biochip industry has grown with miniaturization of the feature size and improvement in analytical capability. In 1995, the first plant microarrays were printed with 96 genes at 200 μm feature size. In 2001, microarrays were prepared that contained 30,000 genes at a feature size of 16 μm. The gene content has increased by a factor of 300 in 6 years. The doubling time is 8 months. The feature size in the microarray industry is headed toward the nanometer range. As the wavelength of optical light is 400 nm, the optical microscopes in vogue may not be sufficient to examine the features of a nanoarray. X-ray scanning devices are needed. Other similarities between microprocessors and microarrays are parallelism, miniaturization, and automation.

Regardless of the type of biomolecule, the NanoPrint™ microarrayer is a robust and customizable platform for all microarray manufacturing applications. High quality, precision microarrays are manufactured by the NanoPrint™ systems. They utilize TeleChem's ArrayIt brand, which is patented and widely used, Professional, 946 and Stealth Style Micro Spotting Pins. Superior linear drive motion control technology and proprietary Warp1 controllers from Dynamic Devices are used by NanoPrint Systems. Compatibility with all standard microarray surfaces made by ArrayIt and other vendors are high for NanoPrint systems. Easy configuration is allowed for by NanoPrint systems to print microarrays into the flat bottoms of 96 well plates. This is enabled by taking advantage of its flexible deck configuration and easy to use software interface. The Microarray-Manager Software features include a method creation wizard, user and version control management, custom calibration of the slide and micro plate positions, complete sample tracking, support of input/output data files, custom array designs, speed profiles and wash protocols, automatic method validation, runtime sample and spotting views, as well as a simulation mode and easy to use graphical reprint wizard. Superior instrument performance in both speed and precision can be seen in the high-speed, high-precision linear servo control system of the NanoPrint. It also has efficient bench top design, user configurable worktable, humidity and dust control, a host of available options, and flexible and sophisticated software.

By measurement of gene expression levels as a function of cell and tissue type and storage in databases, a deeper insight into multicellular development can be achieved. The human brain tissue has been actively studied to date but other tissues such as liver, breast, prostate, lung, colon, kidney, heart, bladder, and skin have not been studied. The key to longevity causing genes can be obtained from nanoarray methods. The onset and progression of human disease are determined by a complex set of factors: genetics, diet, environment, and presence of infectious agents. Oncological studies, diabetes, cardiovascular disease, Alzheimer's disease, stroke, AIDS, cystic fibrosis, Parkinson's disease, autism, and anemia are under intense investigation using microarray analysis by scientists around the world. One goal of microarray analysts is to eradicate every human disease by the year 2050.

In principle, microarrays can be used for drug discovery and clinical trials by generating gene expression profiles in patients undergoing drug treatment. Many drugs impart their therapeutic action to specific cellular targets, inhibiting protein function and altering the gene expression. Many illnesses result in specific changes in gene expression. Drugs that reverse these changes are expected to ameliorate the disease. The cost of drug development may be reduced. Safer medicines can be produced using nanoarray studies. Microarrays can be used for patient genotyping. Microarrays can be used in genetic screening and diagnostics. Thousands of disease-causing sequence variants are known. Affordable microarray screens for these diseases are of tremendous scientific and commercial interest. The population can be delineated as normal, carrier, and disease genotypes by use of microarray screening. Genetic diseases curable at an early stage can be identified. Health care costs can be reduced by use of genetic testing kits. Commonly inherited diseases such as cystic fibrosis, sickle cell anemia, Tay Sachs disease, and breast cancer can be studied using microarrays. The genomic information can be provided to the public by confidential access.

The polypeptide sequence information can be obtained by use of protein chips. *Metabolomics* is the complete functional annotation of the genome. This is a natural next step to genomics and proteomics. The functions of the organism are governed by the signals from the proteins, which are generated by DNA. Nanoarray analysis can be used in metabolomics.

Gene chip technology is a practical method for measurement of the sequence of genetic building blocks. The search for disease related genetic changes could be speeded up. A team of scientists at Johns Hopkins University was able to accurately determine the order of 2 million blocks of each of 40 individuals' genomes in just a year. This is in a fraction of time required by traditional technology. Only 10 errors out of every 10,000 points were detected. Researchers at Washington University School of Medicine in St. Louis, Missouri, explained how genes dictate our biological clock. The circadian rhythm was studied using microarray analysis. How do you feel when you get up at 4:00 AM in the morning compared with 4:00 PM in the afternoon? Events such as this are driven by our internal clock, connected to external cues like the Sun. Thus far, products of eight different genes have been discovered to be essential to the operations of this clock. Three laboratories in collaboration with Affymetrix have identified 22 genes that appear to be rhythmically regulated by the internal clock of the Drosophila fly. Drosophila melanogaster has 14,000 genes. Microarrays can be used to prepare a comprehensive list of all the active genes in a tissue sample. The fly was exposed to light for 12 h, followed by darkness for 12 h. The cycle continued for a total of 96 h. Genetic analyses were performed on half of the flies at six different time-points on the fifth day. Seventy readings of

14,000 genes were taken and one million individual measurements were completed. Sophisticated computer bases statistical analyses were performed and the team determined that between 72 and 200 of the flies' 14,000 genes showed significant rhythm of gene expression in normal flies living in a daily light-dark cycle. *Oscillating genes* were also detected as were mutant flies.

The chip is imbedded with DNA molecules instead of electronic circuitry. It is designed to probe a biological sample for genetic information that indicates whether the person has a genetic predisposition for certain diseases. A University of Houston scientist has developed a chemical process for building a device that could help doctors predict a patient's response to drugs or screen patients for thousands of genetic mutations and diseases, all with one simple lab test. This is a highly parallel technology. Ten thousand experiments can be performed at once.

Aging of the human retina has been found by researchers to be accompanied by distinct changes in gene expression. Using commercially available DNA slides, a team of researchers directed by Swaroop have established the first-ever gene profile of the aging human retina, an important step in understanding the mechanisms of aging and its impact on vision disorders. In the *Journal of Investigative Ophthalmology and Visual Science*, Swaroop[9] and colleagues show that retinal aging is associated, in particular, with expression changes of genes involved in stress response and energy metabolism. The term gene expression means that in any given cell, only a portion of the genes is expressed or switched on. For example, a person's pancreas and retina have the same genes, but only the pancreas can turn on the genes that allow it to make insulin. Swaroop believes that the findings will help scientists understand whether age predisposes one to changes in the retina that, in turn, lead to age-related diseases. For vision researchers, one of the most pressing disorders is age-related macular degeneration (AMD), a progressive eye disease that affects the retina and results in the loss of one's fine central vision. Microarray technology is an important tool for gene profiling because it allows rapid comparison of thousands of genes, something that was unheard of even a few years ago.

Microarrays are detected using fluorescence probes and a confocal scanning microscope. All microarrays require fluorescence scanning to facilitate reliable imaging of the expression pattern of the genes or the problem at hand. The confocal laser scanner delivers the highest image and data quality. Commercial devices such as ScanArray are used currently. In the future, as the minimum feature size of the microarray dot reaches the nanometer range, x-ray scanners may have to be developed as the wavelength of light is 400 nm. The substrate is in the form of chemically treated glass in the form of a 25×75 mm slide. DNA arrays incorporate samples tagged with multiple fluorescent probes. Differential gene expression leads to a ratiometric approach and renders absolute calibration unnecessary. The glass substrate gives minimal background fluorescence and hence is a good choice for the substrate.

Fluorescence in biological detection is a vast topic and has been described comprehensively elsewhere.[10] Fluorescence light is emitted from a dye or fluorophore, which is illuminated by excitation light. The fluorescence emission wavelength is always longer than the wavelength of excitation light. For example, fluorescein isothiocynate (FITC) exhibits an excitation curve peak at 494 nm and an emission peak at 518 nm. The wavelength difference between the emission and excitation peaks is 24 nm. Typical for most dyes used in microarrays, this wavelength difference is called the *Stokes shift*.

The optical requirements of a detection instrument are as follows:

1. Excitation

2. Emission light collection

3. Spatial addressing

4. Excitation/emission discrimination

5. Detection

Confocal scanners have two focal points (Fig. 6.1) configured to limit the field of view in three dimensions. They image a small area with an aim of point resolution using pixels. The collimated laser beam is reflected from the beam splitter into the objective lens.

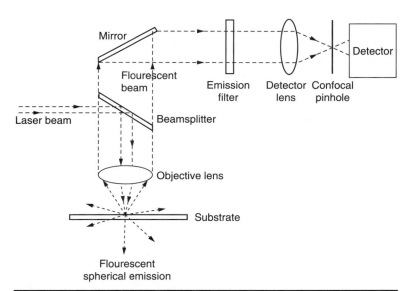

FIGURE 6.1 Confocal scanning arrangement in a microarray scanner.

The laser beam fills only a fraction of the lens. The degree of fill depends on the choice of the lens numerical aperture (NA) and pixel size. The laser beam is focused on the sample where it induces spherical fluorescence in all directions. The excitation beam also reflects back up toward the detector. The objective lens collects a fraction of the spherical fluorescence emission and collimates it into a parallel beam. It also collects the reflected laser light, which is 3 to 7 orders of magnitude higher in intensity compared with that of fluorescent light. The return beam is again directed to the beam splitter, which reflects most of the laser light back toward the laser source and transmits most of the fluorescence beam toward the detector. A mirror then folds the system without any optical functionality followed by the emission filter, which selects a narrow band of fluorescence emission and rejects all remaining laser excitation light. The pinhole arrangement facilitates the depth of focus of the objective lens coinciding with the imaging in the detector.

Restricted depth of focus is a disadvantage of the confocal scanning arrangement. It has a moving substrate scanner. Using a moving lens and a moving substrate, higher light collection efficiencies can be obtained. Useful microarray scanners must detect low levels of fluorescence in the Pico watt range. At these low levels, almost all materials fluoresce—the glass substrate, the chemicals comprising the substrate's surface coating, sample washing chemicals, lenses, filters, and even DNA molecules. The scanning instrument needs to maximize detection of the target dye's emission while minimizing the detection of all the other fluorescence sources. The reflected and scattered light must be rejected even though it is one million times brighter than the dim fluorescent light. A photo multiplier tube (PMT) can detect a single photon or a beam of light low in power. PMT amplifies the photon event into an electron event. By varying the tube high voltage the PMT sensitivity or gain can be varied by a range of 1 to 100. Some of the instrument performance measures are as follows: number of lasers and fluorescence channels, detectivity, sensitivity, crosstalk, resolution, field size, uniformity, image geometry, throughput, and superposition of signal sources.

High-quality surfaces are needed for the preparation of microarray samples. How well the molecules attach to the surface determines the efficiency of the biochemical reactions, the precision of detection, and the quality of the resulting data. A microarray experiment is only as good as the surface used to create it. An ideal microarray surface has to be dimensional, flat, planar, uniform, durable, inert, efficient, and accessible.

There exists an optimal target concentration. This is the number of target molecules per unit volume of printed sample that provides the strongest signal in a microarray assay. Optimal target density is the number of target molecules per unit area on a microarray substrate that provides the strongest signal in a microarray assay. Experiments were conducted and microarray signals are plotted as a function of the target molecule concentration. A 15-base oligonucleotide was printed on a microarray substrate at a concentration range of 1 to 100 μm. Hybridization with probe solution containing a fluorescent 15-mer complementary to the target sequence was performed. The scanning was measured at different target concentrations of 1, 3, 10, 30, 50, and 100 μm. Examination of the results reveals that the fluorescent intensity increases steadily in the range of 1 to 10 μm target and reaches peak intensity at 30 μm oligonucleotide, at which point the signal levels off and decreases significantly as the target concentration reaches 100 μm. At the optimal target concentration, the number of target molecules bound to the microarray surface area can be calculated. Assuming that 30 percent of the printed oligonucleotide couples to the substrate and that a typical printed droplet is 300 pl, a 30-μm solution of oligonucleotide gives 2.6 lakh oligonucleotide molecules μm² of the substrate. This is the optimal target density. Additional calculations reveal that 2.6 lakh molecules per μm² correspond to one oligonucleotide per 400 A² or one target molecule per 20 A in a single dimension. It is interesting that a single stranded DNA is 12 A in diameter. The probe-target duplexes would be approximately 24 A in diameter. Due to major and minor groove, the effective diameter is 20 A. A spacing of one target per 20 A defines the optimal target concentration. More material would cause steric hindrance in the packing. Insufficient target density means too few molecules available for hybridization. Physical interference at higher concentrations causes damage and a fall in the signal intensity. In a similar fashion, *optimal probe concentration* is the number of probe molecules per unit volume of sample that provides the strongest signal in a microarray assay (Fig. 6.2).

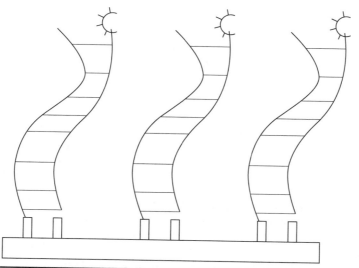

Figure 6.2 Target DNA molecules hybridized with probe molecules with fluorescent tags and attached to the substrate via linker molecules.

Probe concentrations greater than the optimal concentration are useful under certain circumstances. Target (T) molecules on the microarray surface form productive interactions with probe (P) molecules in the solution to form target-probe (T-P) pairs. The generalized biochemical reaction for target-probe binding can be given as:

$$T + P \rightarrow T\text{-}P \tag{6.1}$$

The rate of formation of target-probe products depends on the concentration of the two reactants and can be expressed as the product of the concentration of T and P times a proportionality constant k.

$$\text{Rate} = -k\,(T)(P) = d(T)/dt = d(P)/dt \tag{6.2}$$

As indicated by Eq. (6.2), the reaction between target and probe is a second order biochemical reaction. The constant k is the rate constant. Under optimal experimental conditions, the printed microarray will contain a much larger number of target molecules than are required to form a T-P pair during the course of the reaction. *Target excess* is a kinetic condition in a microarray assay in which the concentration of the target molecules on the surface exceeds the concentration of the probe molecules in solution. Under target excess conditions, the concentration of target molecules is relatively constant and can be lumped with the reaction rate constant term, k. Thus,

$$\text{Rate} = -k'(P) \tag{6.3}$$

k' denotes the fact that the constant target concentration has become part of this term. As can be seen by Eq. (6.3), the reaction rate becomes a pseudo first order expression. Integrating with respect to time:

$$(P)/(P_0) = \exp(-k't) \tag{6.4}$$

The probe molecules are consumed during the course of the reaction in an exponential fashion. Doubling the concentration of a microarray probe solution will double the rate of the reaction. Because faster rates result in more target-probe pairs per unit time and greater (T-P) means greater signal, it is desirable to use as much probe material as possible in any given microarray experiment as long as the performance of the assay is not compromised. The probe concentration that gives the strongest microarray signals is known as the *optimal p*. Saturated condition is reached when most or all of the target elements contain bound probe molecules. Selective target saturation refers to a microarray assay condition in which a subset of the target elements becomes largely or fully bound leading to a loss of quantitation. Signal compression is a microarray assay condition in which the fluorescent readings underestimate the number of molecules present on the target element or in the probe mixture, leading to a loss of assay quantitation.

The glass surface is preferred as the substrate because of the low background fluorescence generated from it. The smoothness of the glass can be measured using a scratch and dig specification. There are different types of glass. The structure of the glass is SiO_2 tetrahedra. The smoothness of the glass surfaces can be accessed at high resolution using AFM. The AFM technique employs a fine silicon tip that traces back and forth across the surface, detecting and recording surface irregularities as it moves. Three-dimensional images are produced in the AFM scans. A typical microarray glass substrate subject to AFM analysis reveals a maximal roughness of 5.3 nm over a 4-μm^2 area, corresponding to a distance of approximately 40 Si-O bonds or about twice the diameter of duplex DNA. Etching refers to a chemical process used to score glass surfaces for the purpose of labeling and identification. The glass surface may be treated by using either amine or aldehyde. Silane reagents are used for this purpose. The reaction of glass with 3-aminopropyltrimethoxysilane is a typical treatment reaction. The overall positive charge of amine microarray surfaces allows attachment of printed biomolecules that carry negative charges. Attachment occurs primarily via electrostatic interactions or attractive forces between positive charges on the amine groups and negative charges on biomolecules such as nucleic acids. Attachment of nucleic acids to an amine surface occurs via interactions between negatively charged amine groups. The DNA phosphate backbone can be attached along the side of the chain with the microarray glass substrate. Denaturation is the process of converting DNA into single strands. Aldehyde surface treatment uses a spacer arm and aminolinker. The substituted amine attaches by covalent coupling. Covalent coupling is an attachment scheme that involves electron sharing between target molecules and the microarray substrate. Molecules couple to an aldehyde surface in a directional manner such that the end of the molecule containing the amino linker bonds to the microarray surface. Proper reaction conditions and blocking agents all but eliminate background fluorescence with aldehyde surfaces. Steric availability is the desirable spatial configuration such as end attachment that maximizes the physical accessibility of target molecules to incoming probe molecules. Blocking agents are chemical or biochemical agent such as borohydrate or bovine serum albumin used to inactivate reactive groups on a microarray substrate to prevent nonspecific reactivity.

Oligonucleotides are short chains of single-stranded DNA or RNA. Single-stranded oligonucleotides provide another common source of target sequences for nucleic acid microarrays. Microarrays of oligonucleotides can be prepared using delivery or synthesis methods. In the delivery strategies, oligonucleotides made offline are prepared using standard phosphoramidite synthesis suspended in a suitable printing buffer and formed into a microarray using a contact or noncontact printing technology. In the synthesis approaches, oligonucleotides are made in situ one base at a time and many synthesis cycles are used until the microarrays are complete. Due to reduced coupling efficiency and large synthesis time, the length of the oligonucleotides is only 5 to 25 nt length. The main advantages of oligonucleotide targets are increased specificity and the capacity to work directly from sequence database information. Two disadvantages of oligonucleotide targets are the requirement for sequence information prior to manufacture and the loss of signal when using certain types of fluorescent probes.

In the early 1980s, Caruthers developed the chemistry used in *phosphoramidite synthesis* in the industry. Phosphoramidite-based oligonucleotide synthesis underlies most of the synthetic DNA market. The DNA market includes 75 commercial vendors world-wide and annual revenues totaling hundreds of millions of dollars. The oligonucleotide of any sequence can be built from the four DNA building blocks. The four DNA bases most often used are known as cyanoethyl phosphoramidites. Each base is identical to its natural counterpart except for the presence of several chemical substituents that protect the phosphoramidites during synthesis and activate the 3′ phosphate for chemical coupling. Three of the phosphoramidite bases, *A*, *C*, and *G*, contain a reactive primary amine on the purine or pyrimidine ring and therefore require a protecting group on the amine to avoid damaging this position during the synthesis process. A benzoyl-protecting group is typically used for bases *A* and *C*, whereas an isobutyryl group is usually employed on *G*. The fourth base, *T*, does not contain a primary amine on the pyrimidine ring and thus does not require a protecting group. All four phosphoramidite bases also contain a dimethoxytrityl (DMT) group on the 5′ hydroxyl, which blocks the 5′ hydroxyl from chemical coupling until it is intentionally deprotected during the synthesis process. Selective deprotection allows synthesis to proceed in a stepwise manner. The 3′ phosphate is protected against side reaction and activated for nucleophilic attack by the presence of β-cyanoethl and diisopropyl groups, respectively. The protecting groups are removed at the end of the synthesis, yielding an oligonucleotide that is identical to native DNA. The synthesis process proceeds in a 3′ and 5′ direction as follows.

The initial step in oligonucleotide synthesis involves coupling the first base to the solid support. Oligonucleotides can be synthesized on a variety of different supports, but the most common matrix is controlled pore glass (CPG). CPG contains pores of identified diameters inside of which synthesis occurs. The deprotection step in oligonucleotide synthesis allows the 5′ hydroxyl to act as a nucleophile attacking the 3′ activated phosphate group of the second base, which is added to the activated CPG matrix by coupling to the first base. The result is dinucleotide bond formation in the 3′ to 5′ direction. After the coupling step, unreacted 5′ hydroxyl groups are inactivated or capped by acetylation to prevent these bases from reacting with phosphoramidites in subsequent coupling steps. Capping prevents

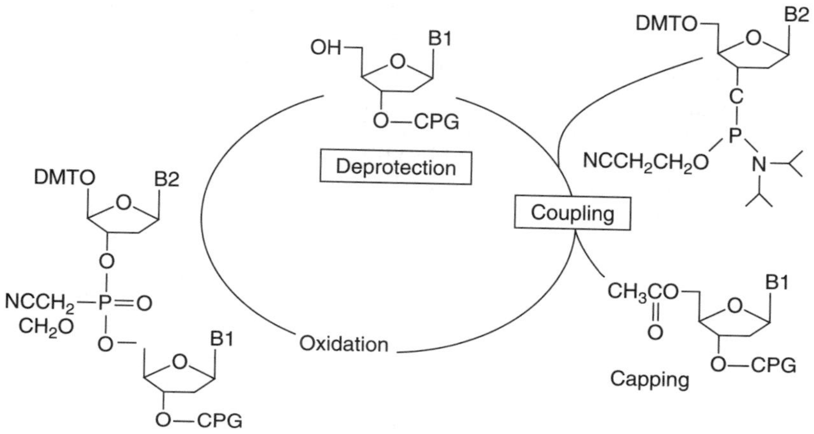

FIGURE 6.3 The four-step process of oligonucleotide synthesis cycle on CPG.

the formation of frame shift oligonucleotides that are missing one or more bases compared to the full-length product, a process that occurs if unreacted 5′ hydroxyls are not capped before the next coupling cycle. After capping, the phosphate trimester of the newly formed dinucleotide is oxidized to the phosphate form to stabilize the phosphate linkage (Fig. 6.3).

The four-step process of deprotection, coupling, capping, and oxidation is the basis of phosphoramidite synthesis and is shown in Fig. 6.3. The oligonucleotide of a known sequence is synthesized by repeating the cycles a few times and using efficiently the right bases and reagents. Each four-step cycle takes 5 to 7 min, enabling synthesis of a synthetic 70 mer in less than 8 h. Following synthesis, the nascent oligonucleotides are treated overnight with ammonium hydroxide to remove the protecting groups from the based and phosphate groups and to cleave the oligonucleotides from the CPG support. With coupling efficiencies exceeding 99 percent per cycle, a synthetic 70 mer preparation would contain more than 60 percent full-length product. Full-length oligonucleotides can be purified away from shorter products using polyacrylamide gel electrophoresis (PAGE) or HPLC.

New superconductors are prepared using a combinatorial mixture of components. A combinatorial synthesis programs is the state-of-the-art mode for discovery of novel drug leads. In addition, in biology, arrays of unique sequences are commonly used to assay the genetic state of cells. In all cases, small volumes of liquids must be precisely metered at high rates of speed. Technology derived from ink-jet printing has been applied to meet such liquid handling needs. Ink-jet printing along with mechanical microspotting and photolithography are the three primary methods of manufacture of microarray slides.

There are two modes of DNA microarray fabrication using ink-jet technology. First is the step-step synthesis of DNA by applying reactive nucleotide monomers to individual surface sites. Second, there is the spotting and immobilization of presynthesized DNA. Ink-jet technology has been used for over 20 years to control delivery of small volumes of liquid to defined locations on two-dimensional surfaces. Different droplet-generating devices are available such as piezoelectric capillary, piezoelectric cavity, thermal, acoustic, continuous flow, etc. Drop diameters of 25 μm can be readily achieved up to 10 kHz by using a piezoelectric device. Smaller diameter droplets and higher frequency than a piezo capillary device can be generated using a piezoelectric cavity device and even higher with a thermal device. No nozzle is used in the acoustic device, which possesses high rates of drop formation (5 mHz) and drop diameters (less than 1 μm). In the continuous flow droplet-generating device, a stream of liquid is broken into distinct droplets by oscillatory pressure. A typical device consists of a static pressure ink reservoir, a small diameter orifice, and a pressure-generating element. The orifice plays a significant role in determining the diameter of the droplets ejected from the device. Drops can be generated on demand and ink can be consumed efficiently. Another method of printing utilizes a continuous stream of droplets directed via an electric or magnetic field onto a print area or alternatively a gutter where the ink is recycled. The printing mode is quite robust as the jet is primed by pressurizing the liquid reservoir. A nozzleless acoustic jet is an interesting development in technology. The drop size can be derived by equating the forces acting on the surface of the drop from the internal and external pressures:

$$\Delta P \,(4\pi) R^2 = \sigma \cdot 2\pi R \qquad (6.5)$$

or

$$R = \sigma/2\Delta P \qquad (6.6)$$

A piezoelectric capillary jet consists of a glass capillary, fixed with an orifice and surrounded by a cylindrical piezoelectric. Droplet formation with piezo capillary jets is accomplished by alternately expanding and contracting the piezoelectric element to generate shock pulses in the fluid chamber. When appropriately tuned to the characteristics of the liquid pressure, pulses sufficient to eject droplets from the nozzle can be generated. Drop formation rates can be up to 10 kHz. Droplet formation is easier at certain frequencies. The size of droplets from these devices depends upon the diameter of the nozzle, the magnitude of the driving force, and the physical properties of the liquid in use. Care must be taken when manufacturing high quality nozzles and in supplying the appropriate waveform in order to obtain droplets that are satellite free and propagating perpendicular to the nozzle plate. Two commercial instruments that are built using this concept are CombiJet and GeneJet.

CombiJet can be used to synthesize DNA microarrays by delivering reagents for phosphoramidite oligonucleotide synthesis to defined locations on glass substrates. GeneJet is used to manufacture DNA microarrays by spotting presynthesized DNA fragments. Localized DNA synthesis is achieved by using jets to deliver reagents for one of two reactions in the phosphoramidite oligonucleotide synthesis cycle. The first uses this single jet device to deliver reagent to deprotect the 5′ hydroxyl position at specific regions on the two-dimensional surface. Oxidation of the phosphor and coupling of one of the four bases is done in bulk chemical treatment of the entire surface. The second method uses five jets, one for each of the four phosphoramidites and one for the activating reagent. CombiJet III was designed to fully automate all steps of DNA microarray synthesis. During in situ synthesis of DNA, no purification is possible. All the reactant products have to remain on the surface. The quality of the material in each locus is thus determined by the stepwise coupling efficiency. In order to evaluate the coupling yield, a set of 64 spots of identical sequence was synthesized with a cleavable attachment to the surface. At the end of 15 cycles, the slide was subjected to a gas phase base reaction to disrupt the surface treatment. The oligos were collected by washing the surface. After complete removal of the remaining protecting groups, the oligo product was end labeled with ^{32}P phosphate and subjected to polyacrylamide gel electrophoresis. The banding pattern of the oligo products was quantitatively analyzed to derive an average stepwise yield of 91 percent. In order to increase efficiency of hybridization and the DNA attachment of the surface, linker molecules can be utilized. Linkers attach themselves to the substrate on one end and to the target molecule on the other. One example of linker molecules is polyethylene glycol polymers. Solvents that are compatible with both the ink-jet hardware and the particular chemical reactions desired are difficult to find. Acetonitrile was used as a solvent for phosphoramidites. For the deprotection reagent, di- or trichloroacetic acid is common. Volatility of these solvents makes them less suitable. Less volatile dibromomethane was used in place of dichloromethane to reduce the loss of solvent during the preparation. This approach offers flexibility and is low in cost. In order to scale up the microarray synthesis into commercial practice, some technical hurdles have to be overcome. With a cycle time of the instrument of 10 min, an array of 18 mer is printed in 3 h. The next generation instrument will be expected to print more than one array at a time. Robustness of jetting has to be improved. Jet-to-jet variability has to be reduced. Sensitivity of drop size to nozzle characteristics needs to be reduced. Change from a uniform glass substrate to a patterned region is desirable.

Deposition of presynthesized biological material is another method of fabrication. The GeneJet III device can use up to eight jets to aspirate samples from 384 or 1536 well microtitre plates and apply them to microarrays. The instrument has five independent axes. Solenoid valves connect the jets. Monitoring by video camera can be used to deliver droplets free from satellites. The equipment is operated in two modes of printing—start-stop mode and print-on-the fly mode. Appropriate software is used in the control of the instrument during its operation. Based on the concentration of material and the expected amount of cross-linking to the surface, 5 to 50 attomoles of material are available in each spot. No shearing of DNA strands has been observed with material up to 2000 base pairs in length. Viscosity limits the length of the DNA that can be studied. At the desired concentration such as $1\ \mu g/\mu L$ DNA of 5 kb and larger will likely be too viscous for a small (30 μm) orifice jet. Viscosity reduction by adding cosolvents may alleviate the problem.

The total time to print a batch of arrays includes the set up time, the time spent cleaning and loading the liquid deposition devices, and the print time itself:

$$\text{Total time} = \text{print time} + \text{fill time} + \text{set up time} \qquad (6.7)$$

All time associated with movement and deposition of spots is included in the print time. Fill time includes all wash steps for the deposition device, time for loading the device, and time for testing the load and getting into position for printing. Set up time includes the time to load the array substrates and microtitre plates containing array element material as well as time to offload the instrument when the batch run is complete. For either the pin or jet instruments, the sum of the print time and fill time can be expressed in terms of the number of instrument cycles and the time per cycle for each component.

$$\text{Print time} + \text{fill time} = (\text{number of cycles})$$
$$\times (\text{print time/cycle} + \text{fill time/cycle}) \quad (6.8)$$

$$T = C(P^*T_{\text{cycle}} + T_f) + T_s \qquad (6.9)$$

where T = total time, C = cycles, P^*T_{cycle} = print time per cycle, T_f = fill time per cycle, and T_s = set up time.

For the pin tool,

$$P^*T_{\text{cycle}} = NT_c \qquad (6.10)$$

where N = number of arrays printed and T_c = contact time per array including motion.

$$C = \frac{G}{P} \qquad (6.11)$$

where G = number of genes and P = number of pins.

Combining terms, the total print time for a batch mode pin device is thus:

$$T_p = \frac{G(NT_c + T_{f,p})}{P} + T_{sp} \tag{6.12}$$

Arrange the arrays as a square; then the number of rows can be calculated:

$$R = \sqrt{N} \tag{6.13}$$

Combining the above terms, the total time for batch printing with the jet approach is:

$$T_j = \frac{G(RT_1 + T_{f,j})}{J} + T_{s,j} \tag{6.14}$$

where R = number of rows of arrays on a platter, T_1 = print time per line, and J = number of jets.

The time for printing for the jet approach depends on the number of rows in the arrays and that of pin printing depends on the number of arrays. It grows linearly with the number of arrays in the pin tool and for the jet instrument changes with the square root of the number of arrays. For equal numbers of jets and pins, the jet instrument will always have the time advantage. The crossover point is independent of the number of genes printed.

6.5 Data Analysis from Nanoarrays

There are many sources of systematic variation in microarray experiments, which affect the measured gene expression levels. *Normalization* is the term used to describe the process of removing such variation, for example, for differences in labeling efficiency between the two fluorescent dyes. In this case, a constant adjustment is commonly used to force the distribution of the log-ratios to have a median of zero for each slide. For cDNA microarrays, the purpose of dye normalization is to balance the fluorescence intensities of the two dyes, green Cy3 and red Cy5, as well as to allow the comparison of expression levels across experiments. Dye bias can be most obviously seen in an experiment where two identical mRNA samples are labeled with different dyes and subsequently hybridized to the same slide. The bias can stem from a variety of factors including physical properties of the dyes (heat and light sensitivity, relative half-life) efficiency of dye incorporation, experimental variability in probe coupling, processing procedures, and scanner settings at the data collection step. The relative gene expression levels measured as log-ratios from replicate experiments may have different spreads due to differences in experimental conditions. Some scale adjustment may then be required so that the relative expression levels from one particular experiment do not dominate the average relative expression levels across replicate experiments.

Using gene expression data from lipid metabolism in mice, Speed developed a normalization procedure. Speed et al. attempted to identify genes with altered expression in apoliprotein AI knock-out mice with low HDL cholesterol levels compared to inbred C57B1/6 control mice. Normalization procedure depends on the experimental setup. Three situations are identified:

1. Within slide normalization

2. Paired slides normalization

3. Multiple slide normalization

A number of considerations influence this decision, such as the proportion of genes that are expected to be differentially expressed in the red and green samples, and the availability of control DNA sequences. Three types of approaches were described:

1. *All genes in the array*: Frequently biological comparisons made on microarrays are quite specific in nature, i.e., only a small proportion of genes are expected to be differentially expressed. Therefore, the remaining genes are expected to have constant expression and so can be used as indicators of the relative intensities of the two dyes. Almost all genes on the array may be used for normalization.

2. *Constantly expressed genes*: Instead of using all genes on the array for normalization, a smaller set of genes called housekeeping genes has constant expression across a variety of conditions, for example, β actin. Although it is very hard to identify a set of housekeeping genes that does not change significantly under any conditions, it may be possible to find sets of "temporary" housekeeping genes for particular experimental conditions.

3. *Controls*: An alternative to normalization by housekeeping genes is use of spike controls or a titration series of control sequences. In the spiked controls method, synthetic DNA sequences or DNA sequences from an organism different from the one being studied are spotted on the array (with possible replication) and included in the two mRNA samples in equal amounts. These spotted control sequences should thus have equal red and green intensities and could be used for normalization. In the titration series approach, spots consisting of different concentrations of the same gene or expressed sequence tag (EST) are printed on the array. These spots are expected to have equal red and green intensities across the range of intensities. Genomic DNA, which is supposed to

have constant expression levels across various conditions, may be used in the titration series. In practice, however, genomic DNA is often too complex to exhibit much signal and setting a titration series that spans the range of intensities for different experiments is technically very challenging.

The apo AI experiment was carried out as part of a study of lipid metabolism and atherosclerosis susceptibility in mice. Apoliprotein AI is a gene known to play a pivotal role in HDL metabolism. The treatment group consisted of eight mice with the apo AI gene knocked-out and the control group consisted of eight normal C57B1/6 mice. For each of these 16 mice, target cDNA was obtained from mRNA by reverse transcription and labeled using a red fluorescent dye, Cy5. The reference sample used in all hybridizations was prepared by pooling cDNA from the eight control mice and was labeled with a green fluorescent dye, Cy3. In this experiment, target cDNA was hybridized to microarrays containing 6384 cDNA probes, including 200 related to lipid metabolism. Each of the 16 hybridizations produced a pair of 16-bit images, which were processed using the software package Spot. The main quantities of interest produced by the image analysis methods are the (R, G) fluorescence intensity pairs for each gene on the array. After image processing and normalization, the gene expression data can be summarized by a matrix X of log-intensity ratios $lg_2(R/G)$ with p rows corresponding to the genes being studied, $n = n_1 + n_2$ columns corresponding to the n_1 control hybridizations (C57BI/6), and n_2 treatment hybridizations (apo AI knock-out). In the experiment considered, $n_1 = n_2 = 8$ and $p = 5548$. Differentially expressed genes were identified by computing t statistics. For genes, j, the t statistic comparing gene expression in the control and treatment groups is

$$t_j = \frac{(x_{2j} - x_{1j})}{\sqrt{\dfrac{s_{1j}^2}{n_1} + \dfrac{s_{2j}^2}{n_2}}} \tag{6.15}$$

where x_{1j} and x_{2j} denote the average background corrected and normalized expression level of gene j in the n_1 control and n_2 treatment hybridizations, respectively. Similarly, s_{1j}^2 and s_{2j}^2 denote the variances of gene j's expression level in the control and treatment hybridizations, respectively. Large absolute t statistics suggest that the corresponding genes have different expression levels in the control and treatment groups. The statistical significance of the results was assessed based on p values adjusted for multiple comparisons.

Global normalization methods assume that the red and green intensities are related by a constant factor; i.e., $R = k \cdot G$ and in practice, the center of the distribution of log-ratios is shifted to zero:

$$\log_2(R/G) - \log_2(R/G) - c = \log_2(R/kG) \tag{6.16}$$

A common choice for the location parameter $c = \log_2(k)$ is the median or mean of the log-intensity ratios for a particular gene set. Global normalization methods are mentioned in the preprocessing steps in a number of papers on the identification of differentially expressed genes in single-slide cDNA microarray experiments. In many cases, the dye bias appears to be dependent on spot intensity as revealed by plots of the log-ratio M vs. overall spot intensity A. An intensity or A-dependent dye normalization method may thus be preferable to global methods. A local A-dependent normalization was performed using the robust scatter plot smoother lowess from the statistical software package R[7].

$$\log_2(R/G) - \log_2(R/G) - c(A) = \log_2[R/k(A)G] \qquad (6.17)$$

where $c(A)$ is the lowess fit to the M vs. A plot. The lowess () function is a scatter plot smoother that was found to perform robust locally linear fits. The lowess () function will not be affected by a small percentage of differentially expressed genes, which will appear as outliers in the M vs. A plot. The user-defined parameter f is the fraction of the data used for smoothing at each point; the larger the f value, the smoother the fit. M vs. A plot amounts to a 45° counterclockwise rotation of the $\log(G)$, $\log(R)$ coordinate system. Within print tip, group normalization is simply a (print tip + A) dependent normalization; i.e.,

$$\log_2\left(\frac{R}{G}\right) - -c_i(A) = \log_2\left[\frac{R}{k_i(A)G}\right] \qquad (6.18)$$

where $c_i(A)$ is the lowess fit to the M vs. A plot for the ith grid only and $i = 1,2,\ldots,i$ represents the number of print tips.

Paired slides normalization applied dye-swamp experiments, two hybridizations for two mRNA samples with dye assignment reversed in the second hybridization. The normalized log-ratios for the first slide are denoted by $\log_2(R/G) - c$ and those for the second slide by $\log_2(R'/G') - c'$. Here c and c' denote the normalization functions for the two slides. These could be obtained by any one amongst the within slide normalization methods described previously. If $c \sim c'$, then

$$\frac{1}{2}\left[\log_2\left(\frac{R}{G}\right) - c - \log_2\left(\frac{R'}{G'} - c'\right)\right] \sim \frac{1}{2}\left[\log_2\left(\frac{R}{G}\right) + \log_2\left(\frac{G'}{R'}\right)\right]$$

$$= \frac{1}{2}\log_2\left(\frac{RG'}{GR'}\right) = \frac{1}{2}(M - M') \qquad (6.19)$$

The relative expression levels for the two slides may be combined without explicit normalization by a procedure referred to as self-normalization. The validity of the assumption can be checked using housekeeping genes or genomic DNA. Given that the dye assignments are reversed in the two experiments, one expects that the normalized log-ratios on the two slides are of equal magnitude and opposite sign; i.e.

$$\log_2\left(\frac{R}{G}\right) - c \sim \log_2\left(\frac{R'}{G'}\right) - c' \tag{6.20}$$

Therefore, rearranging the equation and assuming again that $c \sim c'$ the normalization function c can be given by:

$$c \sim \frac{1}{2}\left[\log_2\left(\frac{R}{G}\right) + \log_2\left(\frac{R'}{G'}\right)\right] = \frac{1}{2}(M + M') \tag{6.21}$$

In practice, $c = c(A)$ is estimated by the lowess fit to the plot of $1/2(M + M') = 1/2\log_2(RR'/GG')$ vs. $1/2(A + A')$ where this time all the genes are used. Global normalization amounts to a vertical translation in an M vs. A plot and does not allow for spatial or intensity dependent dye biases.

6.6   Sequence Alignment and Dynamic Programming

Sequence alignment is a process of lining up two or more sequences to obtain matches among them. Sequence alignment can be used to develop cures for autoimmune disorders, phylogenetic tree construction, identify polypeptide microstructure, during shotgun sequencing, in gene finding, restriction site mapping, open reading frame analysis (ORF) analysis, genetic engineering, drug design, protein secondary structure determination, protein folding, clone analysis, protein classification, etc. An *alignment grading function* is introduced[11] to keep track of the degree of alignments and pick the optimal alignment.

Optimal global alignment of a pair of sequences can be achieved in $O(n^2)$ time using the Needleman and Wunsch dynamic programming algorithm. A dynamic programming table is filled and the optimal alignment falls out of the procedure. The different alignments can be identified using trace back procedures. Penalty and rewards are selected such that when the grade of alignment is greater than zero, the number of characters aligned is greater than the number of mismatched characters in the sequence. When the grade of alignment is zero, then the number of mismatched characters is equal to each other. When the grade of alignment is less than zero, then the number of mismatched characters is greater than the number of matched characters. Semiglobal alignment is obtained by awarding no penalty to end gaps or allowing free end gaps. The development of a dynamic programming algorithm consists of characterization of the structure of an optimal solution, recursive definition of the grade of an optimal solution, computation of the grade of an optimal solution in a bottom-up fashion, and construction of an optimal solution from computed information. The space requirement of $O(n^2)$ can be reduced to $O(n)$ using Hirschberg's dynamic array method. Algorithms for finding the longest common subsequence that is less than quadratic time is presented. The Smith and Waterman algorithm can be used for obtaining the optimal local alignment between a pair of sequences using the dynamic programming method in $O(n^2)$ and $O(n^2)$ space efficiency. The affine gap model can be used to define a penalty for gaps and gap lengths in order to obtain biologically meaningful alignments.

Greedy algorithms can be used for aligning sequences that differ only by a few errors. Miller et al. have developed a method that can guarantee optimality in $O(en)$ time, where e is much less than n and in $O(m+n)$ space required. These are implanted in the Unigene database by National Center for Biotechnology Information (NCBI). Other methods for obtaining sequence alignment include the method of significant diagonals, the heuristic method, approximate alignments, hamming, etc. point accepted mutation (PAM) and block substitution matrix (BLOSUM) matrices are provided and the benefits of using them for alignments are outlined. The methods described were applied to sequences with varying microstructure such as alternating, random, and block distribution. The concept of a super sequence was introduced. The *x-drop algorithm* for global alignment was touched upon. The effect of repeats in a sequence on the dynamic programming procedures is explored. Antidiagonal was defined and *banded diagonal* methods were explored. What will happen when the dynamic programming table is sparse? The implications were explored. The stability of global and local alignment was touched upon. Staircase table, inverse dynamic programming, consensus sequence, sequencing errors, and their ramifications were introduced.

Suffix tree construction and representation of a sequence in a suffix tree was described. The *generalized suffix tree* can be used to represent a set of strings and stores all the suffixes of all the strings. The algorithm for suffix tree construction can be completed in $O(n)$ time and $O(n)$ space. Tandem repeat occurrence can be found in a sequence in $O[n\lg(n)+occ]$ time efficiency. Suffix trees can be used to obtain pairwise sequence alignment. One of the sequences is streamed against another sequence that is stored in a suffix tree. Where the query sequence branches off from the stored tree can be caught. This way all matches between the sequences can be determined.

String algorithms can be used to find patterns P in a text T. Nineteen such algorithms were discussed. The Rabin-Karp algorithm can be executed in $\theta(m)$ preprocessing time and $O[(n-m+1)m]$ matching time. The Knuth-Morris-Pratt algorithm can be completed in $\theta(m)$ preprocessing time and $O(n)$ matching time. The Boyer-Moore algorithm can be performed in $O(m+\sigma)$ preprocessing and $O(n)$ matching time. The finite automaton algorithm can be run in $O(m|\Sigma|)$ preprocessing time and $O(n)$ matching time. Suffix trees can be used in string matching.

They can be tapped in to improve the speed in approximate string matching using dynamic programming. Look-up tables can be constructed in $O(|\Sigma|^w + n)$ time where Σ and w are the alphabet and window sizes, respectively. The Raita algorithm, Shift Oralgorithm, Simon algorithm, Colussi algorithm, Galil and Giancarlo algorithm, Not So Naïve algorithm, Horspool algorithm, Quick Search algorithm, Berry-Ravindran algorithm, Smith algorithm, Reverse Factor algorithm, Turbo Reverse Factor algorithm, Forward DAWG Matching algorithm, McCreights's algorithm for construction of suffix trees, Karkainnen and Sander's algorithm, lazy suffix trees, exact string matching using suffix tree, suffix forest, hash tables, and finding the lowest common ancestor are discussed at the end of chapter as exercise problems. CHAOS, LAGAN, MULTI-LAGAN, Shuffle-LAGAN, F index and pairwise alignment of sequences, GLASS, QUASAR, hash table based tools, AVID, flat trees, and distributed suffix trees are also discussed in end of chapter exercises.

6.7 HMMs and Applications

Markov models were explained in detail.[11] Genome sequence from NCBI is obtained and modeled using geometric distribution and the Markov model. The kth order Markov model is defined. Worked examples in the construction of zeroth order, first order, second order, and third order Markov models are illustrated. The potential for the use of geometric distribution to model DNA sequences is explored. Rabiner's tutorial on hidden Markov model (HMM) is referred to. The three questions in HMM, i.e., evaluation, decoding, and learning, are reviewed. The Markov, stationarity, and output independence assumptions are introduced to keep the problems mathematically tractable. The HMM is characterized completely. The number of operations needed to determine the sequence given the HMM, i.e., the evaluation problem that usually takes time $O(N^T)$ where T is the length of the sequence and N is the number of states, can be completed in $O(N^2T)$ time using the forward algorithm. The Viterbi algorithm with optimal path is discussed. HMM applications such as construction of phylogenetic tree, protein families, wheel HMMs to predict periodicity in DNA, generalized HMM, database mining, multiple alignments, classification using HMMs, signal peptide and signal anchor prediction by HMMs, and the Chargaff parity rule prediction are discussed. The analysis of commercial software SAM, HMMER, HMMPRO, MetaMeme, PSI-BLAST, and PFAM was discussed.

6.8 Gene Finding, Protein Secondary Structure

The *relative entropy site selection* problem is NP complete. Hertz and Stormo[12] presented a Greedy approach to develop an efficient algorithm for the relative entropy site selection problem. Profiles with relative entropy scores lower than d will be discarded. An iterative approach is used in the Gibbs sampling method for the solution to the relative entropy site selection problem. The maximum subsequence problem is a corollary to the problem of finding the coding regions in DNA. Bates and Constable[13] suggested an algorithm that solves the maximum subsequence problem using the principle of recursion. Sharma[14] has shown that the maximum subsequence is found in the deepest branch of the binomial heap. The time taken is $O(n)$. All subsequences are available in the binomial heap. The interpolated Markov model (IMM) is implanted in GLIMMER. Markov models from the first to the eighth order are used in this procedure. Prediction of the translation start site codon, the shine dalgarno (SD) sites problem can be solved by using a t statistic and measurement of statistical significance. OWL and dBEST databases are used in a dictionary-based approach to gene annotation. Pachter treated the problem of sequence alignment and gene finding with a unifying framework. GPHMM is both a generalized HMM and a pair HMM. Manhattan networks and Steiner trees are used in the optimization problem of designing efficient search spaces for HMMs. The Viterbi algorithm search space was reduced from $O(D^4N^2TU)$ by three orders of magnitude. The Pachter algorithm consumes $O(n^3)$ where n is the number of highest scoring pairs. It can be reduced to $O[nlg(n)]$ at the expense of increasing the bound for the size of the network from twice optimal to four times optimal. The spliced alignment algorithm for similarity-based gene recognition is reviewed. The Las Vegas algorithm provides the option of "no answer."

The protein secondary structure—α helix, β-sheet, and γ-coil—given the primary sequence of protein can be accomplished using neural networks. Although the number of proteomes sequenced rapidly increases with time, the number of known secondary structures is not commensurate with the proteomes' growth. Empirical correlations have been developed between protein primary structure and protein secondary structure such as in Chou and Fasman rules. Pioneering work on predicting protein secondary structure using *neural networks* was that of Qian and Sejnowski. They used 106 proteins in their study. The fundamentals of neural networks were reviewed. Rost and Sander came up with the most important performance improvement by using evolutionary conformation in their PhD server to predict protein secondary structure. The NN had 819 input units, one hidden and one output layer. The output layer had three units. The work of Riis and Krogh is a redesign of NN architecture to solve the overfitting problem. The NN was designed with a larger input layer, balanced training sets, 160 adjustable parameters for an α-helix network, 300 to 500 adjustable parameters contained in the β-sheet or coil network, ensembles of networks, and filtering for improved prediction. HMMs can be used to predict protein secondary structure. Baldi and Pollastri used DAG-RNNs for protein secondary structure prediction. It is a three-step process. Weights of bidirectional recurrent neural network (BRNN) architecture including the weights in the recurrent wheels can be trained in a supervised fashion. Native subcellular localization of a protein is important for the understanding of gene/protein function.

6.9   Drug Delivery Applications

Materials can be constructed with molecular-size precision. The potential intersection between nanotechnology and life sciences is vast. Some biological units have nanoscale dimensions. These are viruses, ribosome, molecular motors, and components of the extracellular matrix. Many educational programs in the nation have begun to incorporate advances in nanotechnology and life sciences in their curriculum.[15] Transient transport phenomena principles find increasing application in drug delivery systems. The challenge of drug delivery is liberation of drug agents at the prescribed time in a safe and repeatable manner to the intended target site. Sophisticated and potent drugs have been developed in recent years in the biotechnology and pharmaceutical industries. The agents are proteins or DNA. Traditional methods of drug delivery are ineffective because of toxicity in concentration spikes. Effective and toxic regimes are limited. Some of the challenges of drug delivery that need to be hurdled yet are: (1) continuous release of agents over an extended time, (2) local delivery of agents at predetermined rates to local sites such as solid tumors, and (3) improved ease of administration. Controlled-release systems can provide continuous drug release by zeroth order kinetics. The blood levels of drugs remain constant throughout the period of delivery. Miniaturization has lead to improved diagnostic tools such as biosensors, microneedles for transdermal delivery, bioadhesive microdevices for oral delivery, microfluidic delivery systems, implantable immunoisolating biocapsules, and microchips for controlled release.

Engineered devices at the nanoscale can interact directly with subcellular compartments and probe intracellular events. Biological hypotheses can be tested using nanoscale manipulations. Novel drug delivery systems and imaging systems, nanoprobes, etc. can be developed.

Polymer drug delivery systems have been developed that are nanoscopic. These can be injected or inhaled and can be internalized by human cells. Nanoscale delivery systems may be accomplished by self-assembling systems based on liposomes or micelles. The synthetic materials already used in drug delivery systems can be miniaturized. Degradable copolymers of lactide and glycolide can be used. Polymer particles can be injected for circulation or used in local drug release. The drugs can be encapsulated and assembled into a nanoparticle. It has been reported that 300-nm particles can be synthesized with functional DNA within the solid matrix (Fig. 6.4).

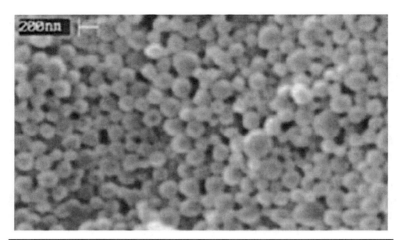

Figure 6.4 Polymer nanoparticle.

Better understanding of the assembly of biocompatible and degradable polymers with the different classes of drug molecules is needed. One of the technical hurdles in this area is to match the particle formation procedure with the drugs. The compatibility between nanostructuring operations and drugs are important. Nanoparticles made from ceramics and minerals offer interesting options to the pharmacist. Cells can internalize nanoparticles. High drug doses used in chemotherapy can be delivered into the cell interior. Conjugation with ligands of polymer nanoparticles can be used to improve the site specificity of the drug targets. Drug carriers smaller than the cell are prepared and large doses of these drugs are delivered directly into the cell's internal machinery. Some recent developments in research indicate that these particles can be made to respond to external stimuli. This can enable remote control of the drug carriers during therapy. By virtue of their smallness, the nanoparticles can offer some interesting alternatives to the physician treating their patient. Novel devices are expected from advances in nanotechnology, enabling interesting applications in biology and medical science. This is because of their combination of subcellular size and controlled release capability and susceptibility to external activation. Dendrimers can be used to improve the efficiency of drug delivery.

Vanderbilt University[16] patented nanoparticulate systems for drug and antigen delivery, gene delivery, and antisense RNA and DNA oligonucleotide delivery. The nanoparticles in the size of 50 to 500 nm can be formed from a variety of synthetic polymers and biopolymers such as proteins and polysaccharides. Nanoparticles can be used as carriers for drugs, antigens, genes, and antisense oligonucleotides. Nanoparticles are formed in a mixture with molecules to be encapsulated within the particles from synthetic or natural polymers using a variety of methods such as phase separation, precipitation, solvent evaporation, emulsification, and spray drying. Nanoparticles can be prepared from copolymers of lactic and glycolic acid or from polyalkylcyanoacrylates. *Nanoencapsulation* of living cells can be affected by polymer precipitation, gelling, or complexing polymer. The inner core material is made of a polyanionic nature and the particle membrane is made up of a combination of polycation and polyanion. The core material is atomized into droplets and collected in a receiving bath containing a polycationic polymer solution. The reverse situation where the core material is polycationic and the receiving bath is polyanionic is also possible.

The nanoparticles are formed in a stirred reactor. The reactor is filled with a cationic solution. A mist of anionic droplets was generated by means of a hollow ultrasonic probe and introduced into the cationic solution in a batch mode resulting in a nonstoichiometric complex with an excess of cationic charge on the particle periphery. Insoluble nanoscopic particles are formed instantaneously. This can be seen by light scattering techniques using the Tyndall effect. The reaction time can be optimized. For particle maturation, usually 1 to 2 h are sufficient. This can be achieved using thermodynamic instability, large surface area, and surface energy. The composition and combinations of anionic polymer mixture as well as that of a cationic receiving bath are essential to allow for adjustments in particle size, shape, and uniformity. Droplets on the converse can be prepared from a polycationic solution and receiving bath containing a polyanionic solution. The ratio of polyanionic and polycationic solutions used is 1:1 to 1:4 and the solution is stirred for 5 to 10 min. Spontaneous formation of nanoparticles was observed for several polymer solutions. The core and corona polymeric solutions are allowed to react in a tubular reactor in the form of mist generated by two separated hollow ultrasonic devices. The product is readily separated because of large differences between the particle and air densities. For the formation of nanoparticles, 33 combinations of polymers can be used.

In summary, the method for preparation of nanoparticles comprises formation of a polyanionic/cationic complex, where the nanoparticles do not dissolve in physiological media for at least one day and is useful in drug delivery. The two polyanionic polymers are contacted with at least two cations. A mist of droplets is captured that contains polyanionic polymers in a cationic liquid and another mist of droplets is captured that contains cations in a liquid containing polyanionic polymers. The nanoparticles have a polyanionic core and polyanionic/cationic complex shell with an excess positive charge on the particle periphery. Polyanionic polymers comprise an anionic antigen or protein resulting in said antigen or protein being incorporated into the polyanionic/cationic complex formed upon contact of polyanionic polymers with said cations. A nonionic polymeric surface modifier is included as a steric stabilizer.

6.10 Summary

Molecular computing is expected to emerge when the limits of minia-turization are realized in silicon chips as the key to further increases in computing speed. DNA molecules and enzymes and biochemical reactions can be used to realize faster operations compared even with a transistors-packed silicon chip microprocessor. They can be used to store and process information. DNA computing started with the molecu-lar algorithm to solve the Hamiltonian path problem. NP complete problems can be treated. Molecular machines can be devised using the better understanding of stereochemistry. Construction of molecular machines is driven by: progress in organic synthesis, powerful computa-tion techniques, and the advent of single molecule analytical tools. Well-defined architectures can be obtained by self-assembly. SAMS, Langmuir-blodgett films, soft lithography, AFM, STM, XPS, etc. can be used to manufacture molecular machines. When designing molecular architectures, the second law of thermodynamics should not be violated. Rotaxane molecules are used to create molecular shuttles, molecular switches, molecular nanovalves, molecular muscles, molecular rotors, and surfaces with controlled wettability. Supramolecular materials offer alternatives to top-down miniaturization and bottom-up fabrication. Self-organization principles hold the key.

Gene expression studies can be carried out in biochips. Target biological materials are examined using fluorescent probes in glass slides packed with thousands of genes. Disease states can be better understood using biochips, and cures from better drug design can be affected. The microarray industry is expected to grow in a similar fashion to the microprocessor revolution. A microarray is an ordered array of microscopic elements in a planar substrate that allows for specific binding of gene or gene products. In order to qualify as a microarray, the analytical device must be ordered, microscopic, pla-nar, and specific. Microarrays are evolving into nanoarrays with the dot size decreasing to the nanoscale. One goal of microarray analysis is to eradicate every human disease by the year 2050. Some of the interesting applications of biochips lie in the areas of gene expression, drug delivery, genetic screening and diagnostics, gene profiling, understanding mechanism of aging, and the study of cancer.

The confocal scanning microscope can be used in microarray detection where fluorescence scanning is used. A laser beam excites the sample, and fluorescence light is emitted from the probe in the sample and can be detected using the difference in wavelength of 24 nm between excitation and emitted light beams. Epi-illumination is used in the scanning process. The excitation and emitted beams pass through the objective lens to and from the sample but in opposite directions. PMT is used as a detecting element. The instrument performance measures are number of lasers and fluorescence channels, detectivity, sensitivity, crosstalk, resolution, field size, uniformity, image geometry, throughput, and superposition of signal sources. High-quality surfaces are needed for the preparation of microarray samples. An ideal microarray surface has to be dimensional, flat, planar, uniform, inert, efficient, and accessible.

Optimal target concentration occurs at a spacing of one DNA target molecule per 20 A. The probe duplex is approximately 24 A. Optimal probe concentration is the number of probe molecules per unit volume of sample that provides the strongest signal in a microarray assay. Microarrays of oligonucleotides can be prepared by using delivery or synthesis methods. The four steps in the process of oligonucleotide synthesis are deprotection, coupling, capping, and oxidation. The three manufacturing methods used during microarray manufacture are ink-jet printing, mechanical microspotting, and photolithography. Stepwise coupling efficiency can be defined to gauge the quality of microarray synthesis. Linker molecules can be used to increase the efficiency of hybridization and DNA attachment at the surface. The time taken for ink-jet printing when jets or pins are used is compared against each other.

Statistical normalization procedures can be used to remove systematic variations in nanoarray experiments that affect the measured gene expression levels. Speed developed a normalization procedure using gene expression data from lipid metabolism in mice. He used housekeeping genes that have constant levels of expression across a variety of conditions. Differentially expressed genes were identified by computing the t statistics. Global normalization methods, M vs. A plot, paired-slide normalization, within slide normalization, and multiple-slide normalization methods are discussed.

Sequence alignment can be used to develop cures for autoimmune disorders, in phylogenetic tree construction, identify polypeptide microstructure, in shotgun sequencing, during drug design, in protein secondary structure determination, in protein folding, clone analysis, protein classification, etc. Optimal global alignment and local alignment can be obtained using dynamic programming. Speedups can be achieved using the Greedy algorithm for nearly aligned sequences. A dynamic array can be used to cut the space required from $O(n^2)$. PAM and Blossum matrices provide different penalties that are specific to the sequences aligned. Sharma discussed the x-drop algorithm, banded diagonal methods, sparse tables, staircase tables, super sequence, inverse dynamic programming, stability of alignment, suffix tree construction, generalized suffix tree procedures, and the advantages of using them. String algorithms can be used to find patterns P in a text T. Nineteen such algorithms were discussed.

Markov models are discussed for varying orders. Three topics in HMM, i.e., the evaluation, the decoding, and the learning, were reviewed. The speedup obtained using the forward algorithm, backward algorithm, and Viterbi algorithm were clarified. Gene finding logarithms were touched upon. Advances made in protein secondary structure prediction were traced from Chou and Fasman rules, to Qian and Sejnowski's neural networks, to the PHD server of Rost and Sander where evolutionary information was used to effect improvement in prediction accuracy. HMMs, DAG-RNNs, and BRNN can be used for protein secondary structure prediction.

The role of polymer nanoparticles in drug delivery applications was discussed. Some of the challenges in drug delivery are continuous release of agents over extended periods of time, local delivery of agents at predetermined rates to local sites such as tumors, and improved ease of administration. Polymer drug delivery systems can be nanoscopic. Self-assembled liposomes and micelles can accomplish the task. Drugs can be encapsulated in the polymer particles. Nanostructuring operations need to be compatible with the drugs. Nanoencapsulation of living cells can be affected by polymer precipitation, gelling, and complexing polymer.

Review Questions

1. What are programmable molecular computing machines?

2. Why is molecular computing expected to be speedier than silicon chip microprocessors?

3. How are ATP molecules and DNA used to devise better storage devices?

4. How did the field of molecular electronics emerge?

5. What is meant by DNA computing?

6. How is DNA computing used to treat NP complete problems?

7. How is the charge distribution in DNA used in the design of molecular computers?

8. What is the importance of autonomous self-assembly of molecular structures?

9. What is meant by aqueous computing?

10. How is the write operation affected in plasmids using restriction enzymes?

11. What is the role of stereochemistry in the design of molecular machines?

12. What are the three factors that gave impetus to construction of molecular machines?

13. What is the relevance to molecular machines of Pople and Kohn winning the Nobel Prize?

14. What are some nanofabrication methods that can be used in construction of molecular machines?

15. Why is the Feynman ratchet a violation of the second law of thermodynamics?

16. How is the host-guest assembly in calixarene molecules used in the construction of molecular machines?

17. What is the role of thermal activation and noncovalent interaction in the design of a molecular shuttle?

18. How are conformers used in the design of molecular switches?

19. What happens to a molecular muscle upon external stimulation?

20. How are controlled molecular motions used in the creation of molecular rotors?

21. What are molecular rotors?

22. What is meant by surfaces with controlled wettability?

23. Write short notes on: (a) molecular nanovalves, (b) supramolecular gatekeepers, and (c) biological nanovalves.

24. What is the role of self-assembly in creation of supramolecular structures?

25. What are the three themes of design and investigation of preorganized molecular receptors?

26. What is meant by: (a) adaptive chemistry, (b) Darwinistic chemistry, and (c) evolutive chemistry?

27. How many transistors per silicon chip and how may genes per biochip can be studied? What is the benefit of such an analogy?

28. What are the four criteria to qualify as a microarray?

29. What are some of the issues involved in the emergence of nanoarrays?

30. What are oligonucleotides?

31. Why is glass preferred as a microarray substrate?

32. Discuss some of the performance parameters of the NanoPrint microarrayer.

33. Can a confocal scanning microscope be used for NanoPrint microarray studies? Why or why not?

34. What are some of the applications of microarray studies?

35. What are the advantages of using microarrays over macroarrays?

36. What are the advantages of using nanoarrays over microarrays?

37. What is meant by protein chips?

38. How is the biological clock studied using microarrays?

39. What is meant by oscillating genes?

40. How is aging of the human retina studied using microarray analysis?

41. What is meant by Stokes shift?

42. What are the five optical requirements of a detection instrument?

43. How many focal points does a confocal scanner have?

44. What are some of the instrument performance measures of a confocal scanning microscope?

45. Can the confocal scanning microscope shown in Fig. 6.1 do without the mirror?

46. Derive the optimal target concentration.

47. Derive the optimal probe concentration.

48. What are some of the advantages of using epi-illumination?

49. What are housekeeping genes?

50. How is the student distribution and t statistics useful in data normalization?

51. What are the differences between delivery and synthesis methods of oligonucleotide synthesis?

52. Discuss the four-step process of the oligonucleotide synthesis cycle.

53. How is combinatorial synthesis used in the design of novel superconductors?

54. What are the differences between mechanical microspotting and ink-jet printing methods of manufacture of microarrays?

55. Comment on the photolithography method of manufacture of microarray slides.

56. What is meant by stepwise coupling efficiency?

57. What is the role of linker molecules?

58. What is the difference between global normalization and paired slides normalization?

59. What are some biological units that possess nanoscale dimensions?

60. What are some of the technical hurdles in drug delivery applications?

61. What is meant by nanoencapsulation?

62. What are microneedles used for in transdermal delivery?

63. Discuss the nanoparticulate system for drug and antigen delivery patented by Vanderbilt University.

64. What is the role of ionic differences between solution and droplets introduced into the stirred reactor?

65. How are polyanionic polymers used in drug delivery design?

66. How are the nanoparticles stabilized upon formation?

References

1. L. M. Adleman, Molecular computation of solutions to combinatorial problems, *Science,* 266, 1021–1024, 1994.
2. H. Yoshida and A. Suyama, Solution to 3-SAT by breadth first search, in *DNA Based Computers 5,* Vol. 54, E. Winfree and D. K. Gifford, Eds., American Mathematical Society, Providence, RI, 1999, pp. 9–20.
3. T. Head, M. Yamamura, and G. Gal, Aqueous computing: Writing on molecules, CEC 99, Proc. 1999 Congress on Evolutionary Computation, Washington, DC, July 1999.
4. J. M. Lehn, Toward self-organization and complex matter, *Science,* 295, 5564, 2400–2403, 2002.
5. R. Kornberg, The molecular basis of eukaryotic transcription, in *Les Prix Nobel,* Stockholm, Sweden: Nobel Committee, 1968.
6. D. Baltimore, RNA dependent DNA polymerase in virions of RNA tumour viruses, *Nature,* 226, 1209–1211, 1970.
7. A. M. Maxam and W. Gilbert, A new method for sequencing DNA, *Proc. of National Academy of Sciences,* 67, 921–928, 1970.
8. F. Sanger and H. Tuppy, The amino-acid sequence in the phenylalanyl chain of insulin. The investigation of peptides from enzymatic hydrolysates, *Biochem J.,* 49, 481–500, 1951.
9. S. Yoshida, B. M. Yashar, S. Hiriyanna and A. Swaroop, Microarray Analysis of Gene Expression in the Aging Human Retina, *Investigative Ophthalmology Visual Science,* 43, 2554–2560, 2002.
10. M. G. Ormerod, Flow Cytometry: A Practical Approach, Oxford University Press, Oxford, UK, 1994.
11. K. R. Sharma, *Bioinformatics: Sequence Alignment and Markov Models,* McGraw-Hill Professional, New York, 2008.
12. G. T. Hertz and G. D. Stormo, Identifying DNA and protein patterns with statistically significant alignments of multiple sequences, *Bioinfomatics,* 15, 563–577, 1999.
13. J. C. Bates and R. L. Constable, Proofs as Programs, *ACM Trans. Program.* Lang. Syst., 7, 113–136, 1985.
14. K. R. Sharma, Binomial tree representation of maximum increasing subsequence problem, 41st Annual Convention of Chemists, Delhi University, N. Delhi, India, 2004.
15. M. Saltzman and T. Desai, Drug delivery in the BME curricula, *Ann. Biomed. Eng.,* 34, 270–275, 2006.
16. A. Prokop, Micro-Particulate and Nano-Particulate Polymeric Delivery System, US Patent 6,726,934, 2004, Vanderbilt University, Nashville, TN.

CHAPTER **7**

Biomimetic
Nanostructures

Learning Objectives

- What are biomimetic materials?
- Discuss equilibrium reactions of self-assembly
- Preparation of hydroxyapetite and collagen-based nanocomposites
- Understand the iridescence of insects and structural colors of plants
- Mechanism of biomineralization in mollusks
- Film formation and properties
- Design of protein scaffold/biomimetic materials and smart materials
- Synthesis of magnetic pigment with nanoparticles and use in memory storage

---

7.1   Overview

Biomimetic according to the dictionary means a compound that *mimics* a biological material in its structure or function or a lab protocol to initiate a natural chemical process. A group of engineers at the University of Florida reported testing a plate that mimics a unique bone in a horse's leg. This was done with an eye toward creating lighter, stronger materials for planes and spacecrafts. The third metacarpus bone in the horse's leg supports the force conveyed as the animal undergoes locomotion. One side of the bone has a small hole. The structure is weakened by the hole and hence the solid structures break upon application of pressure. When the third metacarpus fractures, the hole is intact. Researchers want to mimic the bone's unique strength for airplane and space applications. One classical mechanisms of failure is hole formation. Airplanes have holes for wiring, fuel, and hydraulic lines. The weakness caused by holes is compensated by increasing the thickness of the material around the holes. The structure of horse bone has been studied around the hole. The resulting information was captured in mathematical models. A computer model was developed that has the capability to mimic the bone's behavior under stress. The bone was configured in a manner that the highest stresses are pushed away from the hole into a region of higher strength. A biomimetic plate with a hole surrounded by different grades of polyurethane foam in order to mimic the compositional structure of the bone near the hole was developed. The researchers tested the plate by placing it across two vertical pillars and weighing it down and compared the results with those from an identical test of a plate with a drilled hole without the foam stabilizer. Twice the weight was needed to break the biomimetic plate. The fracture did not go through the hole as occurred with the plate without the foam stabilizer.

For example, Discher[1] prepared rodlike aggregates referred to as worm micelles. These resemble linear proteins found inside cells such as cytoskeleton filament and outside cells such as collagen fibers. The research in solvent diluted copolymer systems was initiated in the early 1990s. Researchers discovered formation of cell-mimetic sacs or vesicles in aqueous solution using amphiphilic block copolymers.

Molecular assembly in biology uses water a lot. Water constitutes 70 percent of body mass in human anatomy. Protein folding phenomena may be due to a hydrophobic effect. The binding of molecules to proteins can also be explained in this manner. Out of the 20 amino acids that comprise the polypeptide protein molecules, nine are hydrophobic. Cell membranes are made up of lipids that are dual hydrophobic-polar. They are arranged in a segmented fashion. This sequential arrangement is referred to as amphiphilic. Hydrophobicity can be used to prepare templates for biomimetic nanostructures through self-assembly. Synthetic mimics of cell components and their functions rely on energetics, stability, and fluidity properties. Block copolymers are segmented into two different monomer units with sections of the polymer having one or the other monomer repeat unit. Copolymers with block microstructure have been found to self-assemble and organize into periodic nanophases such as arrays of rods and stacks of lamellar sheets. Hydrophobic-hydrophilic interactions form the driving force for the formation of structure. The time average molecular shape of an amphiphile in aqueous solution in the corresponding forms of cylinder, wedge, cone, etc. will determine the morphology of the membrane formed such as spherical or rod. The average molecular shape is a function of the hydrophilic fraction. The solvent effects also have a secondary role. Copolymers with block microstructure that are amphiphilic assemble into worm micelles and polymer membranes.

Polyethylene oxide and polybutadiene (PEO-PBd) copolymer with block architecture is an example of an amphiphilic copolymer. PEO is hydrophilic and PBd is hydrophobic. Self-assembly of these copolymers in water can lead to the formation of polymersomes or vesicles. At certain fractions of hydrophilic component, they form rodlike worm micelles. The relatively higher molecular weight result in the aggregate formation. Lipid vesicles are formed into different morphologies such as starfish, tube, pear, or string of pearl shapes. Membrane thicknesses of 3 to 4 nm were confirmed using cryo-TEM. The temperature range of stability of these structures is 273 to 373 K. Flexural Brownian motion and thermal bending modes are important considerations during analysis of stability of the morphologies formed. Membrane bending resistance has been found to increase with increase in polymer molecular weight. The copolymer molecular weight can have a strong effect on vesicle stability and in-plane hydrodynamic properties. Worms less than 10 nm in diameter have been observed using fluorescent labeling.

Diblock copolymer vesicles in aqueous solution have been studied. Protein folding stability is an interesting application of the study. Polyisocyano-L-alanine amphiphile on self-assembly becomes immunogenic. These are used in biomedical applications. The vesicular shells on collapse are 10 to 100 nm in diameter. They coexist with rodlike filaments as well as chiral super helices. Polymersomes can be formed instantaneously by addition of water to lamellar structures of films. Addition of $CHCl_3$, chloroform solutions of copolymer, into water creates vesicles. The chloroform can be removed by dialysis. Cross-linking the PBd can improve the stability of the worms. These systems can be applied in the cosmetics and pharmaceutical industry or as an anticancer agent.

Thus, synthetic polymer vesicles can mimic many biological membrane processes. Examples of such processes are protein integration, protein fusion, DNA encapsulation, and DNA compatibility.

The materials in life sciences can be organized into a hierarchical structure. The levels of structural hierarchy can be seen in a tendon as follows.[2] The fiber diameters are: (1) 0.5 nm for collagen polypeptide, (2) 1.5 nm for triple helix, (3) 3.5 nm for microfibril, (4) 10 to 20 nm for subfibril, (5) 50 to 500 nm for fibril, (6) 50 to 300 μm for fascicle, and (7) 0.1 to 0.5 μm for tendon. Wood and diarthroidial joints are found to have six levels of structural hierarchy. Thin films can be developed using biomimetics. The sequential adsorption of materials onto the surface observed during biomineralization can be mimicked. In the literature such film forming techniques are referred to under different names such as: fuzzy nanoassemblies, polyion multilayers, alternate polyelectrolyte thin films, molecular deposition, bolaform amphile multilayers, polymer self-assembly adsorption, layered composite films, stepwise assembly, and electrostatic self-assembly.

7.2   Equilibrium Kinetics of Self-Assembly

Some polymers have the capability to self-assemble to form complex structures. Some examples of self assembly during thin film formation by the sequential adsorption of materials onto a surface include coil to helix formation, formation of biotin-streptavidin complex, S layer protein formation on a two-dimensional array, and antigen-antibody interactions. A first principle based model of the linear assembly process leading to filaments has been discussed in the literature:

$$2A \Leftrightarrow A_2 \tag{7.1}$$

This dimerization step is a nucleation process. The equilibrium constant for the reversible dimerization reaction can be written as:

$$K_2 = \sigma K \tag{7.2}$$

A linear polymer is formed by addition of monomer to the filament. The recurrent propagation step can be written as:

$$A_{i-1} + A \Leftrightarrow A_i \tag{7.3}$$

The equilibrium constant for the propagation step can be written as:

$$K_i = K \qquad (\text{for } i \geq 3) \tag{7.4}$$

σ is the cooperativity parameter. For small values of this parameter, the subsequent propagation steps are thermodynamically favorable, i.e., $K \gg 1$. During any self-assembly process, the total monomer + aggregate concentration shall be varied (C_0). A certain critical concentration can be derived, C_{crit}. The monomer concentration shall remain less than the critical concentration at all times. For values less than the critical concentration, the monomer concentration increases with C_0. Above the critical concentration only the aggregate formation occurs. The dimer formation step is slow during self-assembly. The propagation steps are faster. The kinetics takes a sigmoidal growth curve.

An example of a self-assembly process can be found in tropomyosin systems that belong to the KMEF family, i.e., keratin-myosin-epidermis-fibrin category of proteins. The repeat pattern in the amino acids is sevenfold, 1234567. The amino acids in positions, 2, 3, 5, 6, and 7 are hydrophilic and the aminoacids in positions 1 and 4 tend to be hydrophobic. This is why the secondary structure of this protein is α-helix. After the formation of the helix conformation, a higher order structure forms on the surface of the helix. The hydrophobic amino acids in positions 1 and 4 form a band on the surface of the helix. The rest of the surface is filled by the hydrophilic amino acids. Two tropomyosin monomers self-assemble into a coiled-coil dimer. The coiled-coil dimer can take on different morphologies such as filaments from head to tail interactions, muscle fibers, etc.

Self-assembly found in proteins has been attempted to be mimicked in synthetic proteins.

7.3 Biomimetic Materials

Natural materials found in the anatomy of living creatures provide bioinspiration for the materials technologist. Structure-function relations in natural tissues are studied. Then the biomimetic materials are engineered. The particularly popular functions in organisms are growth and functional adaptation, hierarchical structuring, damage repair and self-healing, capture of light by eyes, photosensitive erection of plants, wings that enable flight, etc. Galileo is considered to be the father of biomechanics. He reckoned that the shape of an animal's bones is largely adapted to its weight. He observed that the weight of an animal is a function of the cube of its characteristic length. The structural strength of the bone varies proportional to the cross-section or square of the characteristic length. The aspect ratio he deduced will have to decrease with the increase in body weight of the animal. Different approaches of designing a material stems from growth for living beings and fabrication for mimetic materials. A machine component is designed and the material is selected per the knowledge and experience in functional requirements, maximum load during service, fatigue, etc.

One unique property of biomaterials is their capability for self-repair. Investigators have found sacrificial bonds between molecules that break and reform dynamically. For example, during the deformation of wood, bonds were found to undergo reformation and breakage in cycles.[3] This is similar to plastic deformation observed in metal and alloys. *Osteoclasts*, specialized cells in bone, remove material irreversibly and *osteoblasts* deposit material to form virgin tissue. This cycle allows a continuous structural adaptation to external conditions and removal of damaged material by new tissue. A sensor/actuator system is in place that replaces damaged material. The growth direction of a tree post landslide is an example in this regard. A fractured or broken tissue is healed naturally. The mechanism involves formation of an intermediate tissue based on the response to inflammation, followed by scar tissue. Bone tissues are an exception to this empirical observation. They tend to regenerate completely. Much research is underway in self-healing materials and this represents an opportunity for biomimetic materials research.

Biomimetic materials design starts with the observations of structure-function relationships in biomaterials. Systemization of this approach over serendipity is preferred. Cuticle of arthropods was designed to endure IR and ultraviolet (UV) irradiation and the demands of sensory movement, transmission, etc. Biomimetic solutions are stored in large databases. Engineers in search of technical solutions can retrieve these. Biomimetic solutions can be classified according to their functions in the databases. Validation and verification of biomechanisms is an iterative process between life sciences and engineering.

Hydroxyapetite and collagen bone, like nanocomposites, were prepared by Kikuchi et al.[4] A self-organization mechanism between the hydroxyapetite and collagen surfaces was used in the preparation of nanocomposite. The composites prepared were found to possess good biocompatibility and biointegrative activities. They are equivalent to autogenous bone and perform better compared with other synthetic bone materials. These nanocomposites are poised to be used in medical and dental fields in the future. This reduces the patients' loads including pain at the donor sites of autogenous bone after transplantation.

Bone in human anatomy is chiefly composed of hydroxyapetite and collagen. Collagen is a protein abundantly found during the formation of life on the earth but for insects. Extracellular matrices are constructed such as tendons, ligaments, skin, and scar tissue using collagen. Hydroxyapetite is a stable calcium phosphate at a pH of 7.2 to 7.4. They are found in body fluids of vertebrates. They have an affinity for organic molecules. They can be used to filter and separate DNA and proteins. Endoskeletons of vertebrates by evolution have selected collagen and hydroxyapetite as their constituents. Bone is one of the human organs where turnover occurs by metabolism but the mechanical properties of the bone are intact. The turnover process is triggered by attachment of osteoclasts to repaired parts of bone. Hydroxyapetite nanocrystals are dissolved by release of protons from osteoclasts that are attached to bone forming the clear zone, which distinguishes the resorption and other parts on the bone. Collagen fibrils are decomposed by collagenase and other proteases secreted by the osteoclasts. *Howship's lacunae* are cavities created by osteoclastic bone resorption. Osteoblasts cover the surfaces of the lacunae formed. These osteoblasts form the bone via collagen and subsequent calcium and phosphorous release. Hydroxyapetite nanocrystal deposits on the c-axis and a bundle of collagen are formed. This nanostructure plays a salient role in bone metabolism and mechanical properties.

There is a lot of interest in the literature to prepare hydroxyapetite/collagen nanocomposites. Their biocompatibility is tested using implantation techniques. Some of them are self-setting. Hydroxyapetite crystals are grown on collagen fibers using $CaHPO_4$ as precursors of hydroxyapetite. Mimesis of bone nanostructure is required to function as bone in recipient sites.

A detailed description of bone nanostructure formation has not yet been reported. Under healthy conditions, the supersaturated hydroxyapetite and body fluid solution does not deposit on collagen and other organic substances. Presence of ions of Ca^{2+}, PO_4^{3-} may have contributed to calcification of collagen fibrils. Stable formation of hydroxyapetite is under alkaline pH of 8 to 9. Collagen fibrillogenesis occurs at 40°C body temperature. Lengths of hydroxyapetite/collagen fibers grown were 20 µm and those of collagen molecules grown were 300 nm. Electron diffraction patterns of the fibers indicated crescent-like 002 diffraction of hydroxyapetite. The c-axes of hydroxyapetite nanocrystals are aligned along the elongation direction of the hydroxy-apetite/collagen fibers. The orientation is similar to that of the bone of vertebrates. The fibril lengths are based upon the degree of self-organization of hydroxyapetite and collagen as measured by TEM and diffraction patterns. During the alkaline pH, conditions and body temperature promote calcium ion accumulation and the first phase of hydroxyapetite nucleation on the collagen surface. Collagen fibrillo-genesis is promoted by neutral surface charge of collagen achieved by sodium (Na) and chlorine (Cl) ions in physiological saline. With self-organization, the bending strength of the composite was found to increase. Excess water was removed by consolidation. The pH and temperature were sensitive parameters in the determination of bending strength of the resulting nanocomposite.

Hydroxyapetite formation on Langmuir-Blodgett monolayers indicates a driving force for the surface interaction between hydroxy-apetite and collagen as an interfacial interaction between these molecules. This can be deduced from the formation of hydroxyapetite nanocrystals on a carboxyl-terminated monolayer but not on an amino-terminated monolayer. The interfacial interaction was studied using Fourier transform infrared (FTIR) spectrometer using the Kramers-Kronig equation for energy shifts of residues at the interface of hydroxyapetite and collagen. Red shifts in the spectra were found, which meant a decrease in bonding energies of C-O bonds. The hydroxyapetite crystal structure consists of two different Ca sites.

Wistar rats and beagle dogs were used to study the biological reactions. Biocompatibility of the nanocomposite specimens were studied under TEM and SEM. Tissue granulation and surface erosion were observed after 2 to 4 weeks. Collagen fibers encapsulated the debris of composites. Large cells with round nuclei infiltrated into the regions around the composites. Infiltrating macrophages phago-cytize the resulting debris. Composites implanted into the subcutaneous tissue are collapsed from their surfaces. The composites were collapsed continuously and phagocytized for 24 weeks after implantation. Infil-tration of macrophages into the nanocomposites occurred in a similar manner as implanted collagen sponges. Lymphocytes were not observed in both the hydroxyapetite/collagen and collagen sponge implan-tation. This is due to the difference of the rejection, mobilization, and activation of granulocytes. The nanocomposites possess good biocompatibility in comparison with collagen sponges. Bone tissue reaction was examined in SD rates to understand this mechanism. Nanocomposite cylinders were implanted into SD rats. They were observed using optical microscopy after 1, 3, 5, 7, 14, and 28 days. The cut sections stained with hematoxyline and eosin, tartrate-resistant acid phosphatase (TRAP) and alkaline phosphate (ALP) were studied under the microscope. Progressive resorption of the composites was found.

Good osteocompatibility was found after direct bonding between new bone and composites without fibrous connective tissue in the surroundings. Howship's-like resorption was found in stained sections after 14 to 28 days. TRAP activity was raised on day 5 in the surroundings and cracks of composites. ALP activity also showed progress in surroundings of new bone. The substitution process of the composites to new bone occurs as follows:

1. Formation of the composite debris by erosion of body fluid

2. Phagocytosis of the debris and composite surface by macrophages

3. Induction of osteoclastic cells on the composite surface and resorption of the composite by an analogous process to that of the bone

4. Induction of osteoblasts to the resorption lacunae created by osteoclastic cells and formation of new bone in the surroundings of the composite

These steps are similar to autogenous bone transplantation. Reconstruction of a critical bone defect in the tibia of beagles was examined for possible clinical use. The tibia defect was 20 mm in length and was formed by surgical saw using Ilizarov bone fixator. The nanocomposite resorption and bone growth were observed at each week until sacrifice date by soft x-ray photography. The interface between the composite and bone was unclear at 10 to 12 weeks after implantation. Nanocomposite resorption and new bone formation in a beagle's tibia is analogous to bone remodeling in rats. The nanocomposite was studied for the presence of human bone morphogenetic protein 2 to exploit the large surface area of contained hydroxyapetite nanocrystals with high adsorbability to organic substances. The nanocomposite was found to be a useful carrier of the protein.

7.4 Biomimetic Thin Films

The iridescence of insects and structural colors of plants are not well understood. The structure and composition of the chromophores are known. Structural colors are different from pigmented colors. Investigations on optical thin films in biology have been undertaken for decades. Structural colors can be altered with application of pressure, swelling or shrinking, or addition of solvent. Addition of swelling agent can result in a change of color in a reversible manner in iridescent wing membranes. Thin film optical interference can explain this observation. Light scattering causes the white color seen in insect wings. Structural colors can be studied using electron microscopy as seen in butterfly and moth scales. These serve as thin film interference filters. Each scale is a flattened stack with two surfaces. The upper lamina contains a grid consisting of raised longitudinal ridges regularly joined by cross-ribs. The ridges and cross-ribs form a series of windows opening into the scale interior. The ridge structure comprises alternating stacks of high and low refractive index layers. Each ridge acts as a quarter-wave thin film interference mirror with a phase change upon reflection. The optical thickness, nt, of a dielectric stack layer is composed of alternating thicknesses t_a and t_b related by

$$n_a t_a = n_b t_b \qquad (7.5)$$

$$nt = n_a t_a + n_b t_b \qquad (7.6)$$

where nt is the optical thickness of a bilayer composed of a high and low index component. The wavelength of maximum reflection is given by

$$\lambda = 4n_a t_a = 4n_b t_b \qquad (7.7)$$

The wavelength of maximum constructive interference varies from 320 to 348 nm over a wing tilt from 0 to 50°. Two lycaenid butterflies were studied for development of iridescence.[5] Two types of internal reflective structures are closely related by development. The diffraction lattice appears to form within the scale cell boundaries through the assistance of a convoluted series of membranes. Membrane cuticle units are produced that are continuous with invaginations of the plasma membrane. Crystallites are formed that grow toward each other by accretion until the adult morphology arises. Thin film interference laminae are formed from the condensation of a network of filaments and tubes secreted outside the boundaries of the cell. Lattice formation is by self-assembly of material into an FCC Bravais lattice structure. The thin film laminae are formed by stretching of the lattice. The lead reflectance spectrum from *Lindsea lucida* has a blue-green reflection band at 538 nm. Blue fruits of *Elaeocarpus angustifolius* exhibit a reflection band at 439 nm. A multilayer structure within the epidermis consisting of a parallel network of strands 78-nm thick was detected by electron microscopy. The optical thickness was 109 nm and the reflectance maximum was 436 nm.

The mechanism of biomineralization in mollusks has been studied by investigation of "flat pearls." In vivo monitoring is accomplished by placing a glass substrate between the mantle and the inner surface of the mollusk shell. The shell structure contains multiple organic, calcite, and aragonite layers. The process is sensitive to substrate. The crystal phase during nacre formation is controlled by soluble mollusk proteins. The amino acids contained in the proteins were aspirate, glycine, glutamate, and serine residues. The red abalone shell formed the source of the proteins. The composition of the aragonitic composite was studied using gel electrophoresis denaturization. These proteins promote the growth of $CaCO_3$ crystals. *Rhombohedral* calcite morphology was found to form in crystals grown in the absence of soluble protein. *Spherulitic* calcite morphology was found to form in crystals grown in the presence of calcitic protein fraction. *Aragonite needles* are formed in the presence of aragonitic fraction. Calcite to aragonite transition is caused by the addition of aragonitic polyanionic proteins. When soluble aragonite proteins are depleted, sequential transition of calcite to aragonite and back to calcite was caused. AFM was used to study the mechanism of aragonite tablet growth. Iridescent patches of organic material are formed when organic pearls are demineralized. Organic sheets with pore diameter of 5 to 50 nm in diameter were found using AFM and scanning ion conductance microscopy.

Crystal CdS synthesis was demonstrated in films made up of polyethylene oxide (PEO). The factors governing the synthesis are strong binding, solubility of the reagents, ordered, and a regular environment to induce nucleation. The crystals formed were found to be uniform in size, phase, and crystallographic orientation. The morphology type of the crystals formed was rock salt form.

Sequential lying down of inorganic layers forms a critical step in biomineralization in mollusk shell. A positively charged substrate is dipped into an aqueous solution containing polyelectrolytes of the negative charge. The negatively charged polyelectrolyte adsorbs to the surface. Upon rinsing and drying, the film is dipped into a solution containing a positively charged polyelectrolyte. This process is repeated indefinitely with multiple electrolyte solutions. Multilayer thin film formation is an important aspect to this process. Examples of layer formation in albumin/silica, silica/alumina, treatment of fabrics, and multilayers on metals and mica were shown in the literature.

Clean substrates are needed for an efficient film formation. The glass substrates are acid cleaned using concentrated sulfuric acid (H_2SO_4) and hydrogen peroxide (H_2O_2). The substrate surface needs to be modified with charged groups some of the time. Silanol groups are sometimes added. Negative charges are imparted to quartz surfaces. Carboxyl groups are added to gold substrates. The substrate used in polymethylformamide (PMF) film is PET. Amide linkages are formed between carboxylate groups and some polycyclic aromatic hydrocarbon (PAH) amino groups. A net positive charge is affected to silanized slides by dipping into hydrochloric acid (HCl).

Polyion multilayer films are characterized by small angle x-ray reflectance (SAXR) spectroscopy. UV/Vis spectroscopy taps into the chromophore present in PMF. Film thicknesses are measured using ellipsometry. The deposition kinetics as well as film thickness can be measured using surface plasmon resonance spectroscopy. Material deposition can be studied using IR spectroscopy. Real-time monitoring of rate and amount of monolayer deposition during PMF monolayer formation is allowed for by use of quartz crystal microbalance (QCM). QCM is a piezoelectric device. Mass charges on the order of nanograms are quantitatively measured. The changes in resonant frequency of a quartz crystal with the changes of mass of material loaded into the crystal are given by the Sauerbrey equation. The mechanism of PMF film formation can be studied by AFM. Hectorite sheets of 25 to 30 nm were imaged using AFM. The charged macromolecules adsorb onto surface defects at short deposition times. These form islands and retain their coil conformation. Homogeneous monolayers composed of flattened polymer chains are found at longer deposition times. Monolayer formation in PMF can be studied using AFM. Surface coverage is measured as a function of adsorption time and ionic strength. Initial kinetics is diffusion limited and at long times it becomes random sequential adsorption. Surface coverage is sensitive to the ionic strength.

Biopolymers have been prepared such as polysaccharide-containing PMFs. In chitosan/polysaccharide (PSS) film thickness increased with dipping solution ionic strength ranging from 15 A per bilayer in the absence of salt to 69 A in a molar sodium chloride. Ionic strength was found to be an important parameter in determination of adsorption kinetics. Saturation condition was achieved at lower molarity. Shielding of chitosan charge by added salt provided for more conformational flexibility and enabled adsorption at the surface. A film containing alternating layers of DNA and PAH, a polymer/biopolymer hybrid, was prepared. Coil conformation of polypeptides can be detected using circular dichroism. β-sheet conformation formed from self-assembly between two polypeptides can be seen using IR analysis of the film.

Streptavidin-containing films were described in the literature. A precursor was used. The film was irradiated with UV light through a copper mask. The film was then immersed in a solution containing FITC-labeled streptavidin. Fluorescence spectroscopy was used to study the protein arrays. Protein containing PMFs were studied. The molecular weight range of proteins studied is 12,400 to 240,000. A multilayer was prepared consisting of glucose isomerase and the bolamphiphile in porous trimethylamine polystyrene beads. The carrier pore diameter was 46 nm and only two layers of enzyme could deposit on the pores. Enzyme activity was comparable to that of soluble and monolayer enzyme preparations. Enzyme activity was studied in films containing up to 40 enzyme layers. In the 40-bilayer film, the average activity per layer decreased by 50 percent of that measured for a 10-layer film. This is apparently because of the inability of the substrate to diffuse deeply into the film.

Sequential adsorption is a low-cost approach to the assembly of thin films. Most polyions can be incorporated into a film including dyes, polymers, proteins, viruses, inorganic nanoparticles, and ceramic plates. Automation of the technique is possible. Minimal equipment investment is required. Films with interesting features can be synthesized by hand using beakers, stopwatch, water, and electrolytes. Complex multilayers are formed using an automated slide stainer. Scale-up of the sequential adsorption technique is less expensive compared with the Langmuir-Blodgett technique. Some examples of PMFs were discussed[5]. The properties can be tuned by varying the number of layers or the spacing between the layers. Any substrate where a charge can be placed can be used in the synthesis of these films. The sequential adsorption process exhibits self-healing characteristics. Point defects and dust inclusions have less penetration distance. The ionic strength is a sensitive parameter in varying the bilayer thickness. This technique can be combined with other techniques.

Three-dimensional control of film composition and properties are provided. Spin coating comprises application of a solution of film material into a rapidly spinning disk. A uniform fluid film solidifies upon evaporation of the solvent. Solvent casting involves drying of a polymer solution placed in a well. Oriented thin films are formed when cast in magnetic or electric fields. Films have been formed by direct polymerization into an initiator covalently attached to the substrate. In the Langmuir-Blodgett technique, an amphiphile monolayer is placed on an air-water interface. The temperature in the water bath is controlled. The surface pressure is measured and controlled by a Teflon arm touching the interface containing the monolayer. The monolayer is transferred to the glass slide using a mechanical dipping apparatus. Complex optical films can be prepared this way. Practical application of this technique is precluded by high cost and poor efficiency. The control of molecular structure is made possible using photolithography. This is affected by a combined method of solid-phase peptide synthesis and semiconductor-based photolithography.

The sequential adsorption technique for film formation can be applied in LEDs, conducting polymers, second order nonlinear optics (NLO), dye-containing optical film, polydiacetylene, bioreactors, molecular recognition by antibody-antigen interaction, nonthrombogenic surfaces, and nanoscale thin film pH electrode. Starch can be converted to gluconic acid using the sequential adsorption film formation process. The reaction rate achieved is 0.0045 $mol/m^2/h$ compared with 0.017 $mol/L/h$. A 1 m^2 membrane would have approximately one-third the efficiency of a commercial microbial fermenter. A two-component film has different levels of structure hierarchy: (1) monolayer, (2) bilayer, and (3) multilayer. A three-component film has five levels of hierarchy.

7.5 Biomimetic Membranes

A protein scaffold/biomimetic membrane material was developed at Argonne National Laboratory. It is a tool for encapsulating and studying the native behavior and structure of membrane and soluble proteins. The membrane material is a complex fluid made up of a mixture of a lipid, polymer amphiphile, a cosurfactant, and water. It undergoes thermoreversible phase changes and exists as a liquid below a certain threshold temperature and as a liquid crystalline gel above that temperature. Dedicated proteins, enzymes, and other biomolecules are mixed and ordered in the liquid state and ordered by increasing the temperature above room temperature. The orientation of the materials is further increased by use of magnetic fields. When applied to selected substrate materials, domains can be oriented preferentially. Certain nonionic, amphiphilic triblock copolymers of PEO-PPO-PEO can be employed as an alternative to more expensive PED with materials currently in use that are also architecturally limited. They can be used as lipid conjugates for producing biomimetic nanostructures. The polyphenyleneoxide (PPO) chain length when approximately similar to the dimensions of the acyl chain region of the lipid bilayer results in a strongly anchored triblock copolymer.

Medical researchers used biomimetic nanostructures to examine soft tissue cellular wounds such as burns, frostbite, radiation exposure, pressure trauma, electric shop injuries, scrapes and abrasions, heart attacks, and stroke. They can be used as a drug screening and development tool as in the following:

1. Nano band-aids for augmenting healing of cellular wounds
2. Nanocapsules for site-directed delivery of healing agents
3. Ideal polymers for healing soft tissue damage
4. Nerve regeneration in spinal cord injuries

Synthetic biological membranes with self-organizing characteristics such as liquid crystalline gels that change shape and function as a response to environmental changes were developed at the University of Chicago.[6] There is increased interest in development of "smart materials." The properties of these materials change in response to environmental stimuli such as ionic strength, temperature, and magnetic or electric fields. The basis of molecular machines, chemical valves and switches, sensors, and a wide range of optoelectronic materials is the response of bioorganisms to external stimuli.

A mixture of lipids, a low-molecular weight polyethylene glycol-derived polymer lipid, and a pentanol surfactant is one such example of a synthetic biological membrane. These gels transform to liquid by heating to an elevated temperature. At higher temperatures, the incorporated proteins and pentanol surfactant rapidly degenerate. The material undergoes phase separation at lower temperatures. Material responds to external stimulus of temperature alone. The material developed at the University of Chicago was in response to a need for development of materials that are responsive to a variety of external stimuli. The material must also be biocompatible.

The developed material exists as a gel at elevated temperatures and a liquid at lower temperatures. It is used in drug delivery systems where body temperature is high. The developed materials are biocompatible, membrane-mimetic liquid crystalline material. The biologically active membrane proteins can be encapsulated using the developed material within an organized lipid matrix. Macroscopic ordering of molecules occurs when a magnetic field, an electric field, or shear is applied to the mixture. The mixture is multipositional. The material reacts to external stimuli to provide both structural and functional characteristics. It manifests a birefringent phase when subjected to a certain environment, an optically isotropic or transparent phase. An optical cue is given when an intact membrane has formed in response to the application of certain stimuli. The material undergoes a *thermoreversible* phase change. It comprises a lipid, polymer amphiphile such as polymer grafted phospholipids, cosurfactant, and water. When the temperature is increased, the material solidifies.

The material comprises 65 to 90 percent water, 3 to 5 percent surfactant, 7 to 27 percent lipid and amphiphilic polymer, and the ratio of polymer to lipid is approximately 4 to 10 mole percent. The mixture undergoing phase change is depicted in Fig. 7.1. The stimulus responsive fluid developed with self-assembling properties switches between two distinct structural states and two distinct functional states in response to several external stimuli. The material is prepared by noncovalent, self-assembly of a quaternary mixture of phospholipids, a lipopolymer, or diblock or triblock copolymer or polymer grafted amphiphile, and a surfactant dispersed in water. The supra molecular, nondenaturing material undergoes a reversible transformation from a liquid crystalline gel to a nonbirefringent fluid upon reduction in temperature. The liquid phase is found to instantaneously organize into a liquid crystalline gel with an increase in temperature. These changes are at the molecular level but manifest at the macromolecular level. The phase change occurs at a temperature range of 15 to 20°C.

Figure 7.1   Gel undergoing phase change.

The gel composition shown in Fig. 7.1 is a mixture of phospholipids, polymer amphiphile such as end-grafted phospholipids or a diblock or a triblock copolymer, and a zwitterionic or cationic cosurfactant dispersed in water. These gels form bilayer membranes with the hydrophobic ends of the lipid and cosurfactant of each layer oriented inward. Cavities/spaces in these membranes can accommodate the fluid in which the membranes are immersed. When juxtaposed, the cavities/spaces are differentially organized into planar sheets and channels separated by water-impermeable lipid micelle and membranes. Proteins and other substances generally of a size between 1 and 50 nm may be incorporated in the bilayer membranes or in the aqueous channels. This allows membranes to be used in packaging or encapsulation for drug delivery applications. Gels depending on their characteristics can be used as sensors and opto and microelectronic products.

Mesoscopic self-assemblage of the developed fluid is further enhanced when the fluid contacts an appropriate surface. When the gel phase of the mixture interacts with certain surfaces containing OH groups, orientation of lamellar domains of the gel is directed into macroscopic dimensions. This ordering enhancement is because of polar phospholipid head groups contained in the mixture and a similar hydrophilic group on the support substrate. This mechanism enables the mixture to hold target functional groups that are encapsulated by the mixture in a certain orientation.

7.6 Magnetic Pigments

CSIR, India,[7] patented a single-step, cost-effective process for manufacture of *acicular* magnetic iron oxide particles of maghemite phase of size ranging from 300 to 350 nm in the magnetic field at room temperature using a biomimetic method. The product is used as a magnetic memory storage device.

Magnetic medium strength depends on the technique used for making the pigment. Limitations of materials used in magnetic recording are expected to be limited by super paramagnetic relaxation found in nanoscale particles with a top size of 30 nm. Acicular shape is needed for longitudinal recording. The shape anisotropy of the material is characterized by its shape. As this influences the coercive field value, this is a salient consideration. It was difficult to prepare nanoscale particles using the earlier oxyhydroxide method. This was due to poor control over the growth kinetics. Cubic or irregular-shaped particles were formed by oxidation of magnetite particles. Maghemite particles are formed by oxidation of Fe_2SO_4 with KNO_3 followed by addition of NaOH solution slowly at 60 to 80°C with continuous stirring. The black precipitate was dried upon washing and heated to 250°C for 30 min to create maghemite particles.

The CSIR process consists of the following steps:

1. Iron salt solution and polyvinyl alcohol (PVA) are mixed in a ratio of 3:5 at a pH of 2 to 5 and stirred for 20 min using a magnetic stirrer.

2. Heat resulting solution for 24 h in an oven at 30 to 60°C under nitrogen atmosphere to obtain an iron loaded polymer gel.

3. Soak the polymer gel in 2 molar, NaOH solution under a magnetic field of 800 to 1500 G, at temperature 30 to 50°C, for 4 to 6 min.

4. Remove NaCl from the polymer gel and dry under a nitrogen atmosphere for 24 h.

5. Recover pure acicular maghemite particles from dried polymer gel.

The acicular particles prepared are found to have a high aspect ratio. Particle size achieved was in the range of 300 to 350 nm. Polymer matrix provided nucleation and reaction sites in the self-assembled network formed from gelation. The maghemite phase was found to have a tetragonal bravais lattice structure. The particles were uniformly oriented.

This process is performed at room temperature and is a single-step process for production of nanoscale particles. The nanoparticles are agglomeration free with uniform size and shape. They can be used in memory storage devices.

7.7   Biomimetic Sensors

Molecular identification can be performed using biomimetic sensors. Harmful chemicals, bacteria, and viruses can be screened and samples of air, water, and blood can be analyzed. There is increased interest in medical diagnostics, genomics assays, proteomics analyses, drug discovery screening, and detection of chemical warfare agents for homeland security and defense. Clinical diseases and infectious pathogens can be detected. Living systems possess interesting detecting elements such as antibodies, enzymes, and genes. These are called *bioreceptors*. Because of the high specificity of the DNA hybridization process, development of DNA bioreceptor-based analytical elements is of interest in the industry. Biochips were discussed in Chap. 6.

Raman and surface-enhanced Raman scattering (SERS) assay methods including microarrays and biosensors were used in the development of an integrated Raman sensor system.[8] The receptor probes are modified for binding with at least one target molecule. Raman spectroscopy is complementary to fluorescence and can be used as an analytical tool for some applications that require high specificity. In recent years, the Raman technique has been rejuvenated by Raman enhancement by 10^6 to 10^{10} for molecules adsorbed on microstructures of metal surfaces. A technique associated with this phenomena is called SERS spectroscopy. The enhancement is attributed to a microstructured metal surface scattering process which increases the weak normal Raman scattering (NRS) due to a combination of electromagnetic and chemical effects between molecules adsorbed on metal surfaces and the metal surfaces themselves. Enhancement is because of plasmon excitation at the metal surface. The effect is limited to copper, gold, and silver and a few other metals for which surface plasmons are excited by visible radiation. When chemisorption occurs, the Raman signal is enhanced due to the formation of new chemical bonds and the consequent perturbation of adsorbate electronic energy levels leading to a surface-induced resonance effect. Raman spectroscopy offers a plethora of information on molecular structures, surface processes, and interface reactions that can be extracted from experimental data.

A Raman signal is generated by receptor probes when combined with the target molecule sensitive to electromagnetic radiation applied to the target, receptor, or both the receptor and target. An IC-based detection system is used for the detection of the Raman signal from the receptor/target combination. The receptor probes are selected from DNA, RNA, antibodies, proteins, enzymes, cells or cell components, and biomimetics. Biomimetics can be molecular imprint antibodies, DNA-based aptamers, peptide nucleic acid (PNA), cyclodextrins, and dendrimers. The analyzing step can comprise gene mapping, gene identification, DNA sequencing, medical diagnostics, drug discovery screening, and environmental bioremediation. The receptor/target generates a Raman signal upon irradiation with electromagnetic radiation. A sample suspected of containing at least one target is introduced to the receptor. The receptor is then radiated. Raman radiation emanating from the radiating step is then analyzed. Metallic precursor particles are embedded in a polymer gel. The polymer gel is irradiated with light. The nanoparticles are *photogenerated*. The light is directed through a photomask with regions transparent to light. As an aliter, light can be directed through a diffractive optical filter in order to split the light into light beams. Alternatively, the light is directed to a digital microarray mirror as discussed in Chap. 6. A solid support material is provided for coating nanospheres (Fig. 7.2). Then a layer of SERS-active metal is deposited on the nanosphere-coated solid support. This allows for conduction of electrons required for generation of surface plasmons.

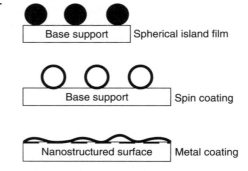

FIGURE 7.2
Surface enhanced
Raman scattering
nanostructure.

Base support | Spherical island film

Base support | Spin coating

Nanostructured surface | Metal coating

7.8 Summary

Biomimetic materials are designed to mimic a natural biological material. For example, the third metacarpus bone in a horse's leg is used as a target for design of aerospace materials. Worm micelles are prepared that resemble linear proteins found in cytoskeleton filament and collagen fibers. Copolymers with block microstructure have been found to self-assemble and organize into periodic nanophases. Molecular shape is found to be a function of a fraction of a hydrophilic fraction. Polymersomes or vesicles can be formed by self-assembly of PEO-PBd in water. Lipid vesicles are formed into different morphologies such as starfish, tube, pear, and string of pearl shapes. Worms with less than 10-nm diameters and membranes with 3-nm thickness have been observed. Stability of protein folding can be studied using self-assembly. Synthetic polymer vesicles can mimic many biological membrane processes.

The equilibrium kinetics of self-assembly reactions was discussed. A cooperativity parameter is defined along with the equilibrium rate constant. A system used as an example to illustrate the mathematical treatment is tropomyosin. Amino acids in positions 1 and 4 are hydrophobic and in positions 2, 3, 5, 6, and 7 are hydrophilic. Banding on helix structures comes about.

One property of biomaterials worthy of mimicking is the ability to self-repair. Biomimetic mechanisms are stored in databases. Hydroxyapetite and collagen were used to prepare bonelike nanocomposite. Howship's lacunae are cavities created by osteoclastic bone resorption. Hydroxyapetite forms on Langmuir-Blodgett monolayers. There is an interfacial interaction between hydroxyapetite and collagen. Substitution process of composites to new bone occurs in stages similar to autogenous bone transplantation: erosion of body fluid and formation of composite debris, phagocytosis of debris, resorption of composite, and induction of osteoblast to the resorption lacunae. Reconstruction of a critical bone defect in the tibia of beagles was examined for possible clinical use.

The iridescence of insects and structural colors of plants are not well understood. Optical thickness of a dielectric stack layer of alternating thickness and wavelength of maximum constructive interference was quantitated. Two lycaenid butterflies were studied for development of iridescence. The mechanism of biomineralization in mollusks has been studied by investigation of "flat pearls." Rhombohedral calcite morphology, spherulite calcite morphology, and aragonite needles are formed under different conditions. Aragonite tablet growth was studied using AFM. Crystal CdS with rock salt morphology was synthesized in films made up of PEO.

Efficient film formation needs a clean substrate. Polyion multilayer films are characterized by SAXR. QCM is used to measure mass charges in nanogram quantity materials. The mechanism of PMF film formation was studied using AFM. Polysaccharide-containing PMF biopolymers have been prepared. Adsorption kinetics depends on ionic strength. A polymer/biopolymer hybrid such as DNA and PAH were formed into a film containing alternating layers. Films containing streptavidin, glucose isomerase, etc. were discussed. Assembly of thin films is by sequential adsorption. Three-dimensional control of film composition and properties was discussed.

Protein scaffold/biomimetic membrane material was discussed. Membrane material is a complex fluid made up of a mixture of a lipid, polymer amphiphile, and a cosurfactant. It undergoes thermoreversible phase changes and exists as a liquid below a certain threshold temperature and a liquid crystalline gel above that temperature. Biomimetic nanostructures are used to examine soft tissue cellular wounds and dry sensing and development and nerve regeneration. Smart materials are developed that undergo a property change in response to environmental stimuli. These materials are used in drug delivery systems. The gel undergoing phase change is shown in Fig. 7.1.

Magnetic pigment used in magnetite memory storage devices with a maghemite phase of size ranging between 300 and 350 nm using biomimetic method was patented by CSIR, India. Longitudinal recording requires acicular shape. Molecular identification can be prepared using biomimetic sensors. High specificity requirements lead to development of the Raman spectroscope. The SERS nanostructure is shown in Fig. 7.2. Nanoparticles can be photogenerated.

Review Questions

1. How is the third metacarpus bone in a horse's leg used in the design of aerospace materials?

2. What is the effect of holes in the horse's bone and the wings of an airplane used for wiring?

3. How much stronger was the biomimetic plate compared with a common plate?

4. Rodlike aggregates referred to as worm micelles were prepared with resemblance to what?

5. What is the role of hydrophobic and hydrophilic properties of protein molecules in protein folding?

6. How are nanophases in copolymers with block microstructures formed?

7. What is the mechanism of formation of polymersomes from a PEO-PBd system?

8. What are the four different morphologies that lipid vesicles can exhibit?

9. What are some of the important considerations during analysis of stability of the morphologies formed?

10. Name some applications of vesicles formed from lamellar structures of films.

11. Discuss the levels of structural hierarchy that can be seen in tendons.

12. What is the equilibrium reaction that can be used to describe the linear assembly process leading to filaments?

13. Under what values of the cooperativity parameter is the polymerization propagation step thermodynamically favorable?

14. What is the connection between aggregate formation and critical monomer concentration?

15. Explain the helix structure formation in the KMEF category of proteins using the repeat pattern in the amino acids and the hydrophobic and hydrophilic properties.

16. Who is considered the father of biomechanics? According to him, what is the relation between the shape of an animal's bone and its weight?

17. Describe the self-repair property of biomaterials.

18. Discuss the osteoclasts and osteoblasts cycle.

19. What is the role of self-organization in the preparation of hydroxyapetite-bone nanocomposites?

20. Explain the properties of biocompatibility and biointegration of synthetic materials.

21. How are hydroxyapetite nanocrystals dissolved?

22. What is meant by Howship's lacunae?

23. Why is dicalcium phosphate used as a precursor to prepare hydroxyapetite?

24. Explain the formation of bone nanostructure in the human anatomy.

25. Explain in vitro fibrillogenesis and the conditions preferred.

26. What are Langmuir-Blodgett monolayers and what is their role in hydroxyapetite formation?

27. Where is the Kramers-Kronig equation used?

28. Why were Wistar rats and beagle dogs chosen for biocompatibility studies?

29. What is osteocompatibility?

30. Discuss the stages involved in the substitution process of composites to new bone structures.

31. Compare the resorption of synthetic nanocomposite with autoegenous bone transplantation.

32. Elaborate on tibia defect.

33. What is the difference between structural colors and pigmented colors?

34. Discuss the mechanism of iridescence in lycaenid butterflies.

35. How is the mechanism of biomineralization studied using "flat pearls"?

36. What are the three morphologies exhibited by aragonite composite?

37. How is crystal CdS synthesis demonstrated in films made up of PEO?

38. Why is SAXR needed to characterize polyion multilayer films?

39. Where is QCM used?

40. How is AFM used to study the mechanism of PMF film formation?

41. What happens to chitosan/polysaccharide film thickness when the solution ionic strength is increased?

42. Discuss the pore formation in glucose isomerase and bolamphiphile multilayers.

43. How is sequential adsorption a low cost approach to assembly of thin films?

44. Explain the Langmuir-Blodgett technique for film formation.

45. Name two applications of biomimetic membranes.

46. Explain the niche property of "smart materials."

47. Under what temperatures are the smart materials in gel form and under what temperatures are they in liquid form?

48. What is meant by a thermoreversible phase change?

49. Discuss the gel phase changes shown in Fig. 7.1.

50. What is the size range of proteins that may be incorporated in the bilayer membranes in the aqueous channels?

51. Explain the phenomena of mesoscopic self-assemblage.

52. What is the size range of maghemite phase in acicular magnetic iron oxide particles?

53. Does the magnetic medium strength depend on the technique used?

54. Under what size range of nanoparticles can the syperparamagnetic relaxation be found?

55. What are the important steps in the CSIR process for magnetic pigments with nanoparticles?

56. What is the significance of the aspect ratio of acicular particles in magnetic pigments?

57. Maghemite phase was found to have what kind of Bravais lattice structure? Discuss.

58. What are bioreceptors and how are they used in the design of biosensors?

59. How is Raman spectroscopy selected for applications that require high specificity?

60. Explain the formation of SERS nanostructure as shown in Fig. 7.2.

61. How are nanoparticles photogenerated?

References

1. D. E. Discher, Biomimetic nanostructures, in *Introduction to Nanoscale Science and Technology*, Springer, New York, 2004, pp. 533–548.
2. D. A. Tirrel, Ed., *Hierarchical Structures in Biology as a Guide for New Materials Technology*, National Academy Press, Washington, DC, 1994.
3. P. Fratzl, Biomimetic materials research: What can we really learn from nature's structural materials?, *J. Royal Soc. Interface*, 4, 637–642, 2007.
4. M. Kikuchi, T. Ikoma, S. Itoh, H.N. Matsumoto, Y. Koyama, K. Takakuda, K. Shinomiya, and J. Tanaka, Biomimetic synthesis of bone-like nanocomposites using the self-organization mechanism of hydroxyapetite and collagen, *Composites Sci. Technol.*, 64, 819–825, 2004.
5. T. M. Cooper, Biomimetic thin films, in *Handbook of Nanostructured Materials and Nanotechnology, Vol. 5, Organics, Polymers and Biological Materials*, H. S. Nalwa, Ed., Academic Press, New York, 2000, p. 711.
6. M. A. Firestone and D. M. Tiede, Synthetic Biological Membrane with Self-Organizing Properties, US Patent 6,537,575, 2003, The University of Chicago, Chicago, IL.
7. A. Sinha, J. Chakraborty, and V. Rao, Single-Step Simple and Economical Process for the Preparation of Nanosized Acicular Magnetic Iron Oxide Particles of Maghemite Phase, US Patent 7,087,210, 2006, Council of Scientific and Industrial Research, New Delhi, India.
8. V. D. Tuan, SERS Diagnostic Platforms, Methods and Systems Microarrays, Biosensors and Biochips, US Patent 7,267, 848, 2007, UT-Battelle, Oak Ridge, TN.

CHAPTER 8

Nanoscale Effects in Time Domain

Learning Objectives

- Six reasons to seek generalized Fourier's law of heat conduction
- Method of relativistic transformation solution to semi-infinite medium Cartesian coordinates
- Method of relativistic transformation solution to infinite medium in cylindrical and spherical coordinates
- Comparison of solution with solution by method of Laplace transforms
- Expressions for penetration distance and thermal lag time
- Three regimes of solution: (1) inertial, (2) Bessel regime, and (3) modified Bessel regime
- Taitel paradox revisited
- Use of final condition in time and solution within second law of thermodynamics
- Expression for relaxation time of materials above which subcritical damped oscillations in temperature may be expected

8.1 Overview

Nanoscale effects in time domain are as important, if not more so, as nanoscale phenomena in space domain in a number of applications. When the advances in microprocessor speed are approaching the limits of physical laws on gate width and miniaturization, there is incentive to reexamine the physical laws at a level of scrutiny never done before. Fourier's law of heat conduction, Fick's law of mass diffusion, Newton's law of viscosity, and Ohm's law of electricity, for instance, are laws derived from empirical observations at steady state.

209

Often times they are used in transient applications and a number of significant deviations between experimental observations and theory have been found. On examination at a molecular level from kinetic theory of gases or free electron theory, it can be shown that these laws result only in the limits of steady state. During transient conditions, more terms have to be accounted for. In this chapter, the nanoscale effects in time domain in heat conduction are discussed.

Recent studies have reported x-ray scattering techniques that capture conformational changes of proteins at the nanosecond time scale using a synchrotron-based structural analytical tool. An enhanced version of WAXS, time resolved wide-angle x-ray scattering (TR-WAXS) is increasingly being used in studies of protein secondary structure. Whereas WAXS time of resolution has been in the order of milliseconds, TR-WAXS increased the time resolution window by six orders of magnitude—to the nanosecond range. Movies of functioning proteins can be generated using TR-WAXS. The photodissociation of carbon monoxide caused conformational changes in hemoglobin protein. The conformational changes of the hemoglobin protein at the nanoscale in the time domain can be viewed by use of TR-WAXS. The folding of cytochrome c and nanosecond conformational changes in myoglobin was shown using TR-WAXS.[1] Efforts are underway to improve the time resolution of TR-WAXS to the picosecond range. Structure changes accompany most protein functions such as binding or release of ligand. These could not be captured prior to the development of TR-WAXS. Kinetics of protein conformational changes can be studied. Detailed structural data are provided by TR-WAXS.

8.2   Six Reasons to Seek Generalized Fourier's Law of Heat Conduction

Fourier's law of heat conduction, Fick's law of mass diffusion, Newton's law of viscosity, and Ohm's law of electricity are physical laws that are used to describe transport phenomena of heat, mass, momentum, and electricity. These phenomenological laws were developed largely from empirical observations at steady states several centuries ago. Although they have been used widely for extended periods, there are a number of applications where massive deviations from theoretical predictions based on these laws have been found. Here are six reasons to seek a generalized Fourier's law of heat conduction:

1. Fourier's law of heat conduction was found to contradict the microscopic theory of reversibility introduced by Onsager.[2]

2. Singularities have been found in the description of transient heat conduction using the Fourier parabolic equations. A "blowup" occurs: (a) at short contact times in the expression for surface flux, for the case of description of transient temperature in a semi-infinite medium subject to constant wall temperature boundary

condition; (b) surface flux for a finite slab subject to constant wall temperature on either of its edges; (c) temperature term in the constant wall flux problem in cylindrical coordinates in a semi-infinite medium solved for by use of the Boltzmann transformation, leading to a solution in exponential integral; and (d) in the short time limit, the parabolic conduction equations for a semi-infinite sphere are solved for by using the similarity transformation.

3. Development of Fourier's law of heat conduction was from observations at steady state and empirical in nature. Its use in a transient state such as at a nanoscale level in the time domain is an extrapolation.

4. Overprediction of theory to experiment has been found in important industrial processes such as fluidized bed heat transfer to surfaces, CPU overheating, adsorption, gel acrylamide electrophoresis, restriction mapping, and laser heating of semiconductors during manufacture of semiconductor devices and drug delivery systems.

5. Landau and Lifshitz[3] examined the solution for transient temperature and noted that for times greater than 0, the temperature is finite at all points in the infinite medium except at infinite location. The inference is that the heat pulse has traveled at infinite speed. However, light is the speediest of velocities. Hence, there is a conflict with the light speed barrier stated by Einstein's theory of relativity. The speed of any mobile object including thermal waves ought to be less than the speed of light.

6. Fourier's law breaks down at the nanoscale space level. This is also referred to as the Casimir limit in some quarters. In this regime, the mean free path of the molecules is greater than the dimension of the object under scrutiny.

In order to better describe transient heat conduction events at the nanoscale level, the damped wave conduction and relaxation equation can be used.[4] A comprehensive insight into the characteristics of the analytical solution using the damped wave conduction and relaxation equation was provided. Maxwell[5] originally suggested this. The equation can be written as follows:

$$q = -k\frac{\partial T}{\partial x} - \tau_r \frac{\partial q}{\partial t} \qquad (8.1)$$

where τ_r is the relaxation time (nanoseconds), q is the heat flux (w/m^2), and k is the thermal conductivity of the medium of conduction ($w/m/k$). Cattaneo and Vernotte postulated this equation in the mid-twentieth century. Reviews have been provided by Joseph and Preziosi.[6] The estimates of the relaxation times are of the order of nanoseconds. Some concerns about the generalized Fourier's law of heat conduction violating the second law of thermodynamics have been expressed. It is going to be shown later in the chapter that investigators who do not use time and space conditions appropriately end up with solutions that appear in violation of the second law of thermodynamics. However, when those conditions are corrected to more physically realistic conditions, well-bounded solutions within the constraints of the second law of thermodynamics can be obtained.

8.3 Semi-infinite Cartesian and Infinite Cylindrical and Spherical Mediums

Consider a semi-infinite medium at an initial temperature of T_0 (Fig. 8.1). For times greater than 0, the surface at $x = 0$ is maintained at constant surface temperature at $T = T_s$, $T_s > T_0$. The boundary conditions and initial condition are as follows:

$$t = 0 \qquad T = T_0 \tag{8.2}$$

$$x = 0 \qquad T = T_s \tag{8.3}$$

$$x = \infty \qquad T = T_0 \tag{8.4}$$

The transient temperature in the semi-infinite medium can be solved for by solving the Fourier parabolic heat conduction equations using the Boltzmann transformation

$$\eta = \frac{x}{\sqrt{4\alpha t}}$$

and shown to be

$$u = \frac{(T - T_0)}{(T_s - T_0)} = 1 - erf\left(\frac{x}{\sqrt{4\alpha t}}\right) \tag{8.5}$$

The heat flux can be written as

$$q^* = \frac{q}{\sqrt{k\rho C_p / \tau_r} (T_s - T_0)} = \frac{1}{\sqrt{\pi\tau}} \exp\left(-\frac{x^2}{4\alpha t}\right) \tag{8.6}$$

$T = T_s$ \hspace{4cm} $T = T_0$

$x = 0$ \hspace{4cm} $x = \infty$

FIGURE 8.1   Semi-infinite medium with initial temperature at T_0.

The dimensionless heat flux at the surface is then given by

$$q_s^* = \frac{1}{\sqrt{\pi\tau}} \tag{8.7}$$

It can be seen that there is a "blowup" in Eq. (8.7) as $\tau \to 0$. For applications with substantial industrial importance such as the heat transfer between fluidized beds to immersed surfaces,[4] large deviations between experimental data and mathematical models based upon the surface renewal theory have been found. The critical parameter in the mathematical models is the contact time of the packets that comprise solid particles at the surface. This contact time is small for gas-solid fluidized beds for certain powder types. Under such circumstances, the microscale time effects may have been significant. The parabolic heat conduction models do not account for these. This is one of the motivations for studying the hyperbolic heat conduction models. It has been shown that the ballistic term in the governing hyperbolic heat conduction equation is the "only" mathematical modification to the parabolic heat conduction equation that can remove the singularity in Eq. (8.7) at short times.

The governing hyperbolic heat conduction equation in one dimension for a semi-infinite medium with constant thermophysical properties, ρ, C_p, k, and τ_r, i.e., the density, heat capacity, thermal conductivity, and thermal relaxation time can be obtained by combining the damped wave conduction and relaxation equation with the energy balance equation to yield

$$\frac{\partial u}{\partial \tau} + \frac{\partial^2 u}{\partial \tau^2} = \frac{\partial^2 u}{\partial X^2} \tag{8.8}$$

where

$$u = \frac{(T - T_0)}{(T_s - T_0)} \qquad X = \frac{x}{\sqrt{4\alpha t}} \qquad \tau = \frac{t}{\tau_r} \tag{8.9}$$

Baumeister and Hamill[7,8] obtained the Laplace transform of Eq. (8.8) and applied the boundary condition at $x = \infty$, given by Eqs. (8.2) and (8.1) to obtain in the Laplace domain

$$\bar{u} = \frac{\exp\left(-X\sqrt{s(s+1)}\right)}{s} \tag{8.10}$$

They integrated Eq. (8.10) with respect to space to obtain

$$H(s) = \int \exp\frac{-X\sqrt{s(s+1)}}{s} dX = -\frac{1}{s\sqrt{s(s+1)}} \exp\frac{-X\sqrt{s(s+1)}}{s} \tag{8.11}$$

The inversion of Eq. (8.11) was obtained from the Laplace transform tables and found to be

$$H(\tau) = \int_0^\tau \exp\left(-\frac{p}{2}\right) I_0 1/2\sqrt{p^2 - X^2}\,dp \tag{8.12}$$

The dimensionless temperature is obtained by differentiating $H(\tau)$ in Eq. (8.11) with respect to X and for $\tau \geq X$

$$u = \frac{\partial H}{\partial X} = -X \int_X^\tau \exp\left(-\frac{p}{2}\right) \frac{I_1 1/2\sqrt{p^2 - X^2}}{\sqrt{p^2 - X^2}}\,dp + \exp\left(-\frac{X}{2}\right) \tag{8.13}$$

Baumeister and Hamill presented their solution in the integral form as shown in Eq. (8.13). In this study, the integrand is approximated to a Chebyshev polynomial and a useful expression for the dimensionless temperature is obtained. This is used to compare the results with the results obtained by relativistic transformation. The dimensionless heat flux can be seen to be

$$q^* = \exp\left(-\frac{\tau}{2}\right) I_0 1/2\sqrt{\tau^2 - X^2} \tag{8.14}$$

The surface heat flux can be seen to be

$$q_s^* = \exp\left(-\frac{\tau}{2}\right) I_0\left[\frac{\tau}{2}\right] \tag{8.15}$$

8.3.1 Chebyshev Economization or Telescoping Power Series

In order to study further the dimensionless transient temperature from the hyperbolic damped wave conduction and relaxation equation, the integral expression given by Baumeister and Hamill in Eq. (8.11) can be simplified using a Chebyshev polynomial. Chebyshev polynomial approximations tend to distribute the errors more evenly with reduced maximum error by use of cosine functions. The set of polynomials, $T_n(r) = \cos(n\theta)$, generated from the sequence of cosine functions using the transformation

$$\theta = \cos^{-1}(r) \tag{8.16}$$

is called Chebyshev polynomials (Table 8.1). Coefficients of the Chebyshev polynomials for the integrand in Eq. (8.11)

$$\frac{I_1 1/2\sqrt{p^2 - X^2}}{\sqrt{p^2 - X^2}}$$

can be computed with some effort. The modified Bessel function of the first order and first kind can be expressed as a power series as follows:

$$\frac{I_1 1/2\sqrt{p^2 - X^2}}{\sqrt{p^2 - X^2}} = \sum_{m=0}^{\infty} \frac{(p^2 - X^2)^m}{4^{2k+1}(m!)(m+1)!} = \frac{\psi^m}{4^{2k+1}(m!)(m+1)!} \tag{8.17}$$

where $\psi = p^2 - X^2$.

$T_0(r) = 1$
$T_1(r) = r$
$T_2(r) = 2r^2 - 1$
$T_3(r) = 4r^3 - 3r$
$T_4(r) = 8r^4 - 8r^2 + 1$
$T_5(r) = 16r^5 - 20r^3 + 5r$
$T_6(r) = 32r^6 - 48r^4 + 18r^2 - 1$

TABLE 8.1 Chebyshev Polynomials

Each of the ψ^m terms can be replaced with their expansion in terms of Chebyshev polynomials given in Table 8.1.

The coefficients of like polynomials $T_i(r)$ are collected. When the truncated power series polynomial of the integrand [Eq. (8.17)] is represented by Chebyshev polynomial (Table 8.2), some of the high-order Chebyshev polynomials can be dropped with negligible truncation error. This is because the upper bound for $T_n(r)$ in the interval $(-1,1)$ is 1. The truncated series can then be retransformed to a polynomial in r with fewer terms than the original and with modified coefficients. This procedure is referred to as Chebyshev economization or telescoping a power series.

Prior to expression of Eq. (8.17) in terms of Chebyshev polynomials, the interval (X, τ) needs to be converted to the interval $(-1,1)$. Therefore, let

$$r = \frac{2\psi - \tau - X}{\tau - X} \quad \text{and} \quad \psi = \frac{r(\tau - X) + (\tau + X)}{2} \tag{8.18}$$

$1 = T_0(r)$
$r = T_1(r)$
$r^2 = \dfrac{1}{2}[T_0(r) + T_2(r)]$
$r^3 = \dfrac{1}{4}[3T_1(r) + T_3(r)]$
$r^4 = \dfrac{1}{8}[3T_0(r) + 4T_2(r) + T_4(r)]$
$r^5 = \dfrac{1}{16}[10T_1(r) + 5T_3(r) + T_5(r)]$
$r^6 = \dfrac{1}{32}[10T_0(r) + 15T_2(r) + 6T_4(r) + T_6(r)]$

TABLE 8.2 Powers of r in Terms of the Chebyshev Polynomials

Further, let

$$\xi = (\tau - X) \quad \text{and} \quad \eta = (\tau + X) \tag{8.19}$$

thus,

$$\psi = \frac{r\xi + \eta}{2} \tag{8.20}$$

Substituting Eq. (8.20) in Eq. (8.17),

$$\frac{I_1 1/2(p^2 - X^2)}{\sqrt{p^2 - X^2}} = \sum_{m=0}^{\infty} \frac{(r\xi + \eta)^m}{2^k 4^{2k+1} m!(m+1)!} \tag{8.21}$$

The right-hand side (RHS) of Eq. (8.21) can be written as

$$\frac{1}{4} + \frac{r\xi + \eta}{256} + \frac{(r\xi + \eta)^2}{49,152} + \cdots + \tag{8.22}$$

A truncation error of

$$\frac{(r\xi + \eta)^3}{18,874,368}$$

is incurred in writing the left-hand side (LHS) of Eq. (8.21) as Eq. (8.23). Replacing the r, r^2, r^3 terms in Eq. (8.22) in terms of Chebyshev polynomials given in Table 8.1 and collecting the like Chebyshev coefficients, T_0, T_1, and T_2, the RHS of Eq. (8.23) can be written as

$$T_0(r)\left(\frac{1}{4} + \frac{\eta}{256} + \frac{\eta^2}{49,152} + + \frac{\xi^2}{98,304}\right)$$

$$+ T_1(r)\left(\frac{\xi}{256} + \frac{2\eta\xi}{49,152} + \right) \tag{8.24}$$

The $T_2(r)$ term can be dropped with an added error of only

$$\frac{\xi^2}{98,304}$$

The order of magnitude of the error incurred is thus

$$O\left(\frac{\xi^2}{98,304}\right)$$

Retransformation of the series given by Eq. (8.24) yields

$$\frac{I_1 1/2\sqrt{p^2 - X^2}}{\sqrt{p^2 - X^2}} = \frac{1}{4} - \frac{X^2}{128} + \frac{\eta^2}{49,152} + \frac{\xi^2}{98,304} + \frac{(p^2 - X^2)}{128} \tag{8.25}$$

The error involved in writing Eq. (8.25) is $4.1 \, 10^{-5} \, \eta\xi$. If Chebyshev polynomial approximation was not used for the integrand and the power series was truncated after the second term, the error would have been $4 \, 10^{-3} r^2$. Substituting Eq. (8.23) in Eq. (8.13) and further integrating the expression for dimensionless temperature,

$$u = \exp\left(-\frac{X}{2}\right) + X\exp\left(-\frac{X}{2}\right)\left(\frac{5}{8} + \frac{X}{16} + \frac{\eta^2}{24{,}576} + \frac{\xi^2}{49{,}152}\right)$$

$$+ X\exp\left(-\frac{\tau}{2}\right)\left(\frac{3}{8} - \frac{\tau}{16} - \frac{X^2}{64} + \frac{\eta^2}{24{,}576} + \frac{\xi^2}{49{,}152}\right) \tag{8.26}$$

It can be seen that Eq. (8.26) can be expected to yield reliable predictions on the transient temperature close to the wave front. This is because the error increases as a function of $4.1 \, 10^{-5} \, \xi\eta$. Far from the wave front, i.e., close to the surface, the numerical error may become significant.

8.3.2 Method of Relativistic Transformation of Coordinates

Sharma[9] developed a relativistic transformation method to solve for the transient temperature by damped wave conduction and relaxation in a semi-infinite medium. A closed form solution for the transient temperature was obtained. The hyperbolic governing equation Eq. (8.8) can be multiplied by $\exp(n\tau)$ and for n one-half reduced to Eq. (8.27) in wave temperature. Thus, the transient temperature was found to be a product of a decaying exponential in time and wave temperature, i.e., $u = W \exp(-n\tau)$. This is typical of transient heat conduction applications. In addition, the damping term in the hyperbolic partial differential equation (PDE) once removed will lead to an equation of the Klein Gordon type that can be examined for the wave temperature without being clouded by the damping component. It can be shown that at $n = 1/2$, the governing equation for temperature, Eq. (8.8) can be transformed as

$$\frac{\partial^2 W}{\partial \tau^2} - \frac{W}{4} = \frac{\partial^2 W}{\partial X^2} \tag{8.27}$$

Equation (8.27) for the wave temperature can be transformed into a Bessel differential equation by the following substitution. Let

$$\psi = \tau^2 - X^2$$

This substitution variable, ψ, can be seen to be a spatio-temporal variable. It is symmetric with respect to space and time. It is for the open interval, $\tau > X$. Equation (8.27) becomes

$$4\psi \frac{\partial^2 W}{\partial \psi^2} + 4\frac{\partial W}{\partial \psi} - \frac{W}{4} = 0 \tag{8.28}$$

Equation (8.28) can be seen to be a Bessel differential equation.[6] The solution to Eq. (8.28) can be seen to be

$$W = c_1 I_0\left(1/2\sqrt{\tau^2 - X^2}\right) + c_2 K_0\left(1/2\sqrt{\tau^2 - X^2}\right) \tag{8.29}$$

It can be seen that at the wave front, i.e., $\psi = 0$, W is finite and therefore $c_2 = 0$. Far from the wave front, close to the surface, the boundary condition can be written as

$$X = 0 \qquad u = 1 \quad \text{or} \quad W = 1 \exp(\tau/2) \tag{8.30}$$

As ψ is a spatio-temporal variable. The constants of integration c_1 can tolerate a function in time. This can be upto an exponential relation in time. Applying the boundary condition at the surface, c_1 can be eliminated between Eqs. (8.29) and (8.30) to yield in the open interval, $\tau > X$,

$$u = \frac{I_0 1/2\left(\sqrt{\tau^2 - X^2}\right)}{I_0(\tau/2)} \tag{8.31}$$

In the domain, $X > \tau$, it can be shown that the solution for the dimensional temperature by a similar approach as above is

$$u = \frac{J_0 1/2\left(\sqrt{X^2 - \tau^2}\right)}{I_0(\tau/2)} \tag{8.32}$$

At the wave front, $\psi = 0$, Eq. (8.28) can be solved and

$$\ln(W) = \frac{\psi}{16} \quad \text{or} \quad W = c_3 \exp\left(\frac{\psi}{16}\right)$$

The temperature at the wave front is thus: $u = c_3\exp(-\tau/2) = c_3\exp(-X/2)$. From the boundary condition at $X = 0$, $c_3 = 1.0$. Thus, at the wave front,

$$u = \exp\left(\frac{-X}{2}\right) \tag{8.33}$$

From Eq. (8.32), the inertial lag time associated with an interior point in the semi-infinite medium can be calculated by realizing that the first zero of the Bessel function, $J_0(\psi)$, occurs at $\psi = 2.4048$. Thus,

$$2.4048^2 = \frac{x_p^2}{\alpha\tau_r} - \frac{t_{lag}^2}{\tau_r^2} \tag{8.34}$$

$$t_{lag} = \sqrt{x_p^2\frac{\tau_r}{\alpha} - 23.132\tau_r^2} \tag{8.35}$$

The penetration distance for a given time instant can be developed at the first zero of the Bessel function. Beyond this point, the interior temperatures can be no less than the initial temperature. Thus,

$$X_{pen} = \sqrt{23.132 + \tau^2} \tag{8.36}$$

The surface heat fluxes for a semi-infinite medium subject to constant wall temperature solved by the Fourier parabolic heat conduction model and the hyperbolic damped wave conduction and relaxation model are compared with each other using an MS Excel spreadsheet. Equations (8.7) and (8.15) are shown side by side in Fig. 8.2. The "blowup" in the Fourier model can be seen at short times. The hyperbolic model is well bounded at short times and reached an asymotic limit of $q^* = 1$ instead of $q^* = \infty$. There appears to be a crossover at $\tau = 1/2$. It was found that for $\tau > 3.8$, the prediction of the hyperbolic model is within 10 percent of the parabolic models. It can be seen from Fig. 8.2 that at large times, the predictions of the parabolic and hyperbolic models are the same. For short times, both qualitatively and quantitatively the predictions of the parabolic and hyperbolic models are substantially different.

It is not clear what happens at $\tau = 1/2$. The hyperbolic governing equation can be transformed using the Boltzmann transformation as follows. Let $\gamma = X/\sqrt{\tau}$. Equation (8.8) becomes

$$-\left(2\gamma \frac{\partial u}{\partial \gamma} + \frac{\partial^2 u}{\partial \gamma^2}\right) = \frac{1}{\tau}\left(\gamma \frac{\partial u}{\partial \gamma} - \gamma^2 \frac{\partial^2 u}{\partial \gamma^2}\right) \tag{8.37}$$

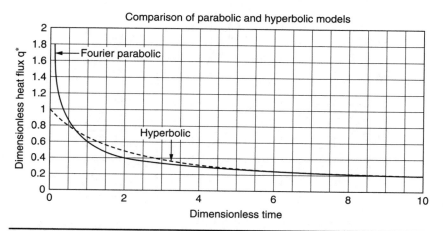

FIGURE 8.2 Comparison of surface flux from the Fourier parabolic heat conduction and hyperbolic damped wave conduction and relaxation models.

For long times, such as $\tau > 1/2$ the RHS of Eq. (8.37) can be dropped and the LHS solved for to yield the solution that is identical with the solution of the Fourier parabolic heat conduction equation, i.e.,

$$u = 1 - erf\left(\frac{X}{\sqrt{4\tau}}\right) \tag{8.38}$$

When differentiated and the expression for flux is obtained at the surface and $X = 0$, it can be seen that both the parabolic heat conduction equation and hyperbolic heat conduction equation predict the same fall of heat flux with time for large times. This is why that beyond $\tau > 1/2$, the predictions of parabolic and hyperbolic models are close to each other as seen in Fig. 8.2. For short times, $\tau < 1/2$, the microscale time effects become important and when neglected give rise to a singularity as can be seen in Fig. 8.2. Therefore, the hyperbolic heat conduction model needs to be used for short time transient applications.

The temperature solution obtained after the Chebyshev polynomial approximation for the integrand in the Baumeister and Hamill solution and further integration is shown in Fig. 8.3. The conditions

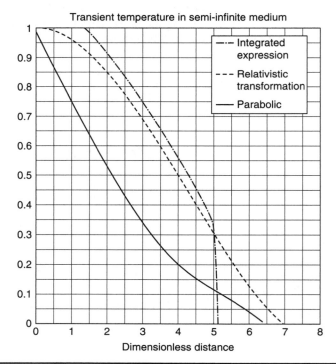

FIGURE 8.3 Temperature distribution in semi-infinite medium by damped wave conduction and relaxation, $\tau = 8$, and parabolic Fourier heat conduction.

selected were for typical $\tau = 5$. Equation (8.25) was plotted using an MS Excel spreadsheet. This is shown in Fig. 8.3. The expression for temperature developed by using the method of relativistic transformation for the same condition of $\tau = 5$ is also shown side by side in Fig. 8.3.

It can be seen that both the Baumeister and Hamill solution and the solution from the relativistic transformation are close to each other, within an average of 12 percent deviation from each other. It can also be seen that close to the surface or far from the wave front the numerical errors expected from the Chebyshev polynomial approximation are large. For such conditions, the expression developed by the method of relativistic transformation may be used. For conditions close to the wave front, the further integrated expression developed in this study may be used. The penetration dimensionless distance for $\tau = 5$ beyond which there is expected no heat transfer is given by Eq. (8.35) and is 6.94 by the method of relativistic transformation.

The Baumeister and Hamill solution is only for $\tau > X$. Both solutions for transient temperature for the damped wave conduction and relaxation hyperbolic equation from the method of Laplace transforms and Chebyshev economization and the method of relativistic transformation are compared against the prediction for transient temperature by the Fourier parabolic heat conduction model. The transient temperature from the Chebyshev economization was found to be within 25 percent of the error function solution for the parabolic Fourier heat conduction model. The hyperbolic model solutions compare well with the Fourier model solution for transient temperature close to the wave front and surface (to within 15 percent of each other). The deviations are at the intermediate values.

8.3.3 Method of Relativistic Transformation of Coordinates in Infinite Cylindrical Medium

Consider a fluid at an initial temperature T_0. The surface of the cylinder is maintained at a constant temperature T_s for times greater than 0. The heat propagative velocity is given as the square root of the ratio of the thermal diffusivity and relaxation time

$$V_h = \sqrt{(\alpha / \tau_r)}$$

The two time conditions, initial and final, and the two boundary conditions are

$$t = 0 \quad r > R \quad T = T_0 \tag{8.39}$$
$$t > 0 \quad r = R \quad T = T_s \tag{8.40}$$
$$r = \infty \quad t > 0 \quad T = T_0 \tag{8.41}$$

The governing equation in temperature is obtained by eliminating the second cross derivative of heat flux wrt r and t between

FIGURE 8.4
Semi-infinite
medium in
cylindrical
cordinates heated
from a cylindrical
surface.

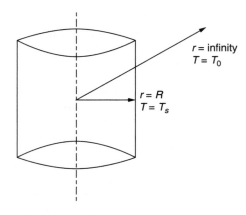

$r = $ infinity
$T = T_0$

$r = R$
$T = T_s$

the non-Fourier damped wave heat conduction and relaxation equation and the energy balance equation in cylindirical coordinates (Fig. 8.4). Considering a cylindrical shell of thickness Δr,

$$\Delta t[2\pi rL\, q_r - 2\pi(r + \Delta r)L\, q_{r+\Delta r}] = [(\rho C_p)\, 2\pi Lr\Delta r\, \Delta T] \tag{8.42}$$

In the limit of Δr, Δt going to zero, the energy balance equation in cylindrical coordinates becomes

$$-\frac{\partial(rq_r)}{r\partial r} = \left(\frac{\rho C_p \partial T}{\partial t}\right) \tag{8.43}$$

The generalized Fourier heat conduction and relaxation equation is

$$q_r = -k\frac{\partial T}{\partial r} - \tau_r\frac{\partial q_r}{\partial t} \tag{8.44}$$

Multiplying Eq. (8.44) by r and differentiating wrt r and then dividing by r,

$$\frac{\partial(rq_r)}{r\partial r} = -\frac{-k}{r}\frac{\partial}{\partial r}\left(\frac{r\partial T}{\partial r}\right) - \frac{\tau_r}{r}\frac{\partial^2(rq_r)}{\partial t\partial r} \tag{8.45}$$

Differentiating Eq. (8.43) wrt t

$$-\frac{1}{r}\frac{\partial^2(rq_r)}{\partial t\partial r} = (\rho C_p)\frac{\partial^2 T}{\partial t^2} \tag{8.46}$$

Substituting Eqs. (8.45) and (8.46) into Eq. (8.43), the governing equation in temperature is obtained as

$$(\rho C_p\tau_r)\frac{\partial^2 T}{\partial t^2} + (\rho C_p)\frac{\partial T}{\partial t} = \frac{k}{r}\frac{\partial}{\partial r}\left(r\frac{\partial T}{\partial r}\right) \tag{8.47}$$

Obtaining the dimensionless variables,

$$u = \frac{(T - T_0)}{(T_s - T_0)} \quad X = \frac{r}{\sqrt{4\alpha t}} \quad \tau = \frac{t}{\tau_r} \tag{8.48}$$

The governing equation in the dimensionless form can be written as

$$\frac{\partial u}{\partial \tau} + \frac{\partial^2 u}{\partial \tau^2} = \frac{\partial^2 u}{\partial X^2} + \frac{1}{X}\frac{\partial u}{\partial X} \tag{8.49}$$

The damping term is removed from the governing equation. This is done realizing that the transient temperature decays with time in an exponential fashion. The other reason for this maneuver is to study the wave equation without the damping term. Let $u = w \exp(-\tau/2)$ and the damping component of the equation is removed to yield

$$\frac{-w}{4} + \frac{\partial^2 w}{\partial \tau^2} = \frac{\partial^2 w}{\partial X^2} + \frac{1}{X}\frac{\partial w}{\partial X} \tag{8.50}$$

Equation (8.50) can be solved by using the method of relativistic transformation of coordinates. Consider the transformation variable η as $\eta = \tau^2 - X^2$, for $\tau > X$. The governing equation becomes a Bessel differential equation for wave temperature,

$$\frac{\partial^2 w}{\partial \eta^2}4(\tau^2 - X^2) + 6\frac{\partial w}{\partial \eta} - \frac{w}{4} = 0 \tag{8.51}$$

$$\eta^2 \frac{\partial^2 w}{\partial \eta^2} + \frac{3}{2}\eta\frac{\partial w}{\partial \eta} - \eta\frac{w}{16} = 0 \tag{8.52}$$

Comparing Eq. (8.52) with the generalized Bessel equation, the solution is

$$a = 3/2 \quad b = 0 \quad c = 0 \quad d = -1/16 \quad s = 1/2$$

The order p of the solution is then $p = 2\sqrt{(1/16)} = 1/2$

$$w = c_1 \frac{I_{1/2}\left(\frac{1}{2}\sqrt{\tau^2 - X^2}\right)}{(\tau^2 - X^2)^{1/4}} + c_2 \frac{I_{-1/2}\left(\frac{1}{2}\sqrt{\tau^2 - X^2}\right)}{(\tau^2 - X^2)^{1/4}} \tag{8.53}$$

c_2 can be seen to be zero as W is finite and not infinitely large at $\eta = 0$. c_1 can be eliminated between Eqs. (8.41) and (8.53). It can be noted that this is a mild function of time, however, because the general solution of PDE consists of n arbitrary functions when the order of the PDE is n compared with n arbitrary constants for ordinary differential equation (ODE). From the boundary condition at $X = X_{R'}$

$$1 = \exp(-\tau/2)c_1 I_{1/2}\left(1/2\sqrt{(\tau^2 - X_R^2)/(\tau^2 - X_R^2)}^{1/4}\right) \tag{8.54}$$

$$u = \left[(t^2 - X_R^2)^{1/4}/(\tau^2 - X^2)^{1/4}\right]$$

$$\times \left[(I_{1/2})\left(1/2\sqrt{(\tau^2 - X^2)}/I_{1/2}\left(1/2\sqrt{(\tau^2 - X_R^2)}\right)\right)\right] \tag{8.55}$$

In terms of elementary functions, Eq. (8.55) can be written as

$$u = \frac{(\tau^2 - X_R^2)^{1/4}}{(\tau^2 - X^2)^2} \frac{\sinh\left(\dfrac{1}{2}\sqrt{\tau^2 - X^2}\right)}{\sinh\left(\dfrac{1}{2}\sqrt{\tau^2 - X_R^2}\right)} \tag{8.56}$$

In the limit of X_R going to zero, the expression becomes

$$u = \frac{\tau}{\sqrt{\tau^2 - X^2}} \frac{\sinh\left(\dfrac{1}{2}\sqrt{\tau^2 - X^2}\right)}{\sinh\left(\dfrac{\tau}{2}\right)} \quad \text{for } \tau > X \tag{8.57}$$

for $X > \tau$,

$$u = \frac{(X_R^2 - \tau^2)^{1/4}}{(X^2 - \tau^2)^{1/2}} \frac{J_{1/2}\left(\dfrac{1}{2}\sqrt{X^2 - \tau^2}\right)}{I_{1/2}\left(\dfrac{1}{2}\sqrt{\tau^2 - X^2}\right)} \tag{8.58}$$

Equation (8.58) can be written in terms of trigonometric functions as

$$u = \frac{(X_R^2 - \tau^2)^{1/4}}{(X^2 - \tau^2)^{1/2}} \frac{\sin\left(\dfrac{1}{2}\sqrt{X^2 - \tau^2}\right)}{\sinh\left(\dfrac{1}{2}\sqrt{\tau^2 - X^2}\right)} \tag{8.59}$$

In the limit of X_R going to zero, the expression becomes

$$u = \frac{\tau}{\sqrt{X^2 - \tau^2}} \frac{\sin\left(\dfrac{1}{2}\sqrt{X^2 - \tau^2}\right)}{\sinh\left(\dfrac{\tau}{2}\right)} \tag{8.60}$$

The dimensionless temperature at a point in the medium at $X = 7$, for example, is considered and shown in Fig. 8.5. Three different regimes can be seen. The first regime is that of the thermal lag and consists of no change from the initial temperature. The second regime is when

$$\tau_{lag}^2 = X^2 - 4\pi^2 \quad \text{or} \quad \tau_{lag} = \sqrt{(X_p^2 - 4\pi^2)} = 3.09 \quad \text{when} \quad X_p = 7 \tag{8.61}$$

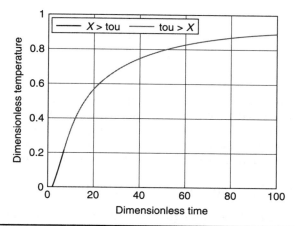

FIGURE 8.5 Transient temperature at a point $X = 7$ in the semi-infinite medium.

For times greater than the time lag and less than X_p, the dimensionless temperature is given by Eq. (8.58). For dimensionless times greater than 7, the dimensionless temperature is given by Eq. (8.55). For distances closer to the surface compared with 2π, the time lag will be zero.

8.3.4 Relativistic Transformation of Spherical Coordinates in an Infinite Medium

Consider a fluid at an initial temperature T_0. The surface of a solid sphere is maintained at a constant temperature T_s for times greater than 0 (Fig. 8.6). The heat propagative velocity is given as the square root of the ratio of the thermal diffusivity and relaxation time, $V_v = \sqrt{(\alpha / \tau_r)}$.

The two time conditions, initial and final, and the two boundary conditions are

FIGURE 8.6
Semi-infinite
medium heated
from a solid
spherical surface.

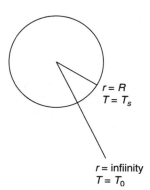

$r = R$
$T = T_s$

$r = \text{infiinity}$
$T = T_0$

$$t = 0 \quad r > R \quad T = T_0 \tag{8.62}$$

$$t = \infty \quad T = T_s \text{ for all } R \tag{8.63}$$

$$t > 0 \quad r = R \quad T = T_s \tag{8.64}$$

$$r = \infty \quad t > 0 \quad T = T_0 \tag{8.65}$$

The governing equation in temperature is obtained by eliminating the second cross derivative of heat flux wrt r and t between the non-Fourier damped wave heat conduction and relaxation equation and the energy balance equation in spherical coordinates. Considering a shell of thickness Δr at a distance r from the center of the solid sphere,

$$\Delta t[4\pi r^2 \, q_r - 4\pi(r + \Delta r)^2 \, q_{r+\Delta r}] = [(\rho C_p) \, 4\pi r^2 \Delta r \, \Delta T] \tag{8.66}$$

Dividing Eq. (8.66) throughout with $\Delta r \Delta t$, and in the limit of Δr, Δt going to zero, the energy balance equation in cylindrical coordinates becomes

$$-\frac{\partial (r q_r)}{r \partial r} = \left(\frac{\rho C_p \partial T}{\partial t} \right) \tag{8.67}$$

The generalized Fourier heat conduction and relaxation equation is

$$q_r = -k\frac{\partial T}{\partial r} - \tau_r \frac{\partial q_r}{\partial t} \tag{8.68}$$

Combining Eqs. (8.67) and (8.68), the governing equation in temperature can be written as

$$(\rho C_p \tau_r)\frac{\partial^2 T}{\partial t^2} + (\rho C_p)\frac{\partial T}{\partial t} = \frac{k}{r^2}\frac{\partial}{\partial r}\left(r^2 \frac{\partial T}{\partial r} \right) \tag{8.69}$$

Obtaining the dimensionless variables,

$$u = \frac{(T - T_0)}{(T_s - T_0)} \quad X = \frac{r}{\sqrt{4\alpha t}} \quad \tau = \frac{t}{\tau_r} \tag{8.70}$$

The governing equation in the dimensionless form can be written as

$$\frac{\partial u}{\partial \tau} + \frac{\partial^2 u}{\partial \tau^2} = \frac{\partial^2 u}{\partial X^2} + \frac{2}{X}\frac{\partial u}{\partial X} \tag{8.71}$$

The damping term is removed from the governing equation. This is done realizing that the transient temperature decays with time in an exponential fashion. The other reason for this maneuver is to study the wave equation without the damping term. Let $u = w \exp(-\tau/2)$ and the damping component of the equation is removed to yield

$$\frac{-w}{4} + \frac{\partial^2 w}{\partial \tau^2} = \frac{\partial^2 w}{\partial X^2} + \frac{2}{X}\frac{\partial w}{\partial X} \tag{8.72}$$

Equation (8.72) can be solved by using the method of relativistic transformation of coordinates. Consider the transformation variable η as $\eta = \tau^2 - X^2$, for $\tau > X$. The governing equation becomes a Bessel differential equation for wave temperature,

$$\frac{\partial^2 w}{\partial \eta^2} 4(\tau^2 - X^2) + 8\frac{\partial w}{\partial \eta} - \frac{w}{4} = 0 \tag{8.73}$$

$$\eta^2 \frac{\partial^2 w}{\partial \eta^2} + 2\eta\frac{\partial w}{\partial \eta} - \eta\frac{w}{16} = 0 \tag{8.74}$$

Comparing Eq. (8.74) with the generalized Bessel equation, the solution is

$$a = 2 \quad b = 0 \quad c = 0 \quad d = -1/16 \quad s = 1/2$$

The order p of the solution is then $p = 1$; $\sqrt{(d/s_r)} = 1/2$

$$w = c_1\frac{I_1\left(\frac{1}{2}\sqrt{\tau^2 - X^2}\right)}{(\tau^2 - X^2)^{1/4}} + c_2\frac{I_{-1}\left(\frac{1}{2}\sqrt{\tau^2 - X^2}\right)}{(\tau^2 - X^2)^{1/4}} \tag{8.75}$$

c_2 can be seen to be zero as W is finite and not infinitely large at $\eta = 0$. c_1 can be eliminated between Eq. (8.64) and Eq. (8.75). It can be noted that this is a mild function of time, however, because the general solution of PDE consists of n arbitrary functions when the order of the PDE is n compared with n arbitrary constants for ODE. From the boundary condition at $X = X_R$,

$$1 = \exp(-\tau/2)c_1 I_{1/2}\left(1/2\ \sqrt{(\tau^2 - X_R^2)}\middle/\left(\tau^2 - X_R^2\right)^{1/4}\right) \tag{8.76}$$

$$u = \left(\frac{\tau^2 - X_R^2}{\tau^2 - X^2}\right)^{1/2}\frac{I_1\left(\frac{1}{2}\sqrt{\tau^2 - X^2}\right)}{I_1\left(\frac{1}{2}\sqrt{\tau^2 - X_R^2}\right)} \tag{8.77}$$

This is applicable for $\tau > X$.
For $X > \tau$, the solution can be written as

$$u = \left(\frac{\tau^2 - X_R^2}{\tau^2 - X^2}\right)^{1/4}\frac{J_1\left(\frac{1}{2}\sqrt{X^2 - \tau^2}\right)}{J_1\left(\frac{1}{2}\sqrt{X_R^2 - \tau^2}\right)} \tag{8.78}$$

Equation (8.78) can be written for $X > \tau$. For $X = \tau$, the solution at the wave front results. This can be obtained by solving Eq. (8.73) at $\eta = 0$. In the limit of X_R going to zero, for $\tau > X$,

$$u = \left(\frac{\tau}{\sqrt{X^2 - \tau^2}}\right) \frac{I_1\left(\frac{1}{2}\sqrt{\tau^2 - X^2}\right)}{I_1\left(\frac{\tau}{2}\right)} \tag{8.79}$$

For $X > \tau$,

$$u = \left(\frac{\tau}{\sqrt{X^2 - \tau^2}}\right) \frac{J_1\left(\frac{1}{2}\sqrt{X^2 - \tau^2}\right)}{I_1\left(\frac{\tau}{2}\right)} \tag{8.80}$$

Seventeen terms were taken in the series expansion of the modified Bessel composite function of the first kind and first order and the Bessel composite function of the first kind and first order, respectively, and the results were plotted in Fig. 8.7 for a given $X_p = 9$ using an MS Excel spreadsheet on a Pentium IV desktop microcomputer. Three regimes can be identified. The first regime is that of the thermal lag and consists of no change from the initial temperature. The second regime is when

$$\tau_{lag}^2 = X^2 - (7.6634)^\wedge 2 \quad \text{or} \quad \tau_{lag} = \sqrt{\left(X_p^2 - 7.6634^2\right)} = 4.72 \text{ when } X_p = 9 \tag{8.81}$$

The first zero of $J_1(x)$ occurs at $x = 3.8317$. The 7.6634 is twice the first root of the Bessel function of the first order and first kind. For times greater than the time lag and less than X_p, the dimensionless temperature is given by Eq. (8.80). For dimensionless times greater than 9, the dimensionless temperature is given by Eq. (8.79). For distances closer to the surface compared with $7.6634\sqrt{(\alpha\tau_r)}$, the thermal lag time will be zero. The ballistic term manifests as a thermal lag at a given point in the medium.

The parabolic Fourier model and hyperbolic model for transient heat flux at the surface for the problem of transient heat conduction

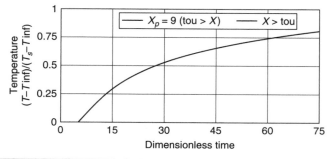

FIGURE 8.7   Transient temperature at a point $X = 9$ in the infinite spherical medium.

in a semi-infinite medium subject to constant surface temperature boundary condition was found to be within 10 percent of each other for times $t > 2\tau_r$ (Fig. 8.3). This checks out with the solution obtained using Boltzmann transformation. The hyperbolic governing equation reverts to the parabolic governing equation during long times. At short times, there is a "blowup" in the parabolic model. In the hyperbolic model, there is no singularity. This has significant implications in several industrial applications such as fluidized bed heat transfer, CPU overheating, gel acrylamide electrophoresis, etc.

The solution developed by Baumeister and Hamill by the method of Laplace transforms was further integrated into a useful expression. A Chebyshev polynomial approximation was used to approximate the integrand with modified Bessel composite function of space and time of the first kind and first order. The error involved in Chebyshev economization was $4.1 \ 10^{-5} \ \eta\xi$. The useful expression for transient temperature was shown in Fig. 8.4 for a typical time of $\tau = 5$. The dimensionless temperature as a function of dimensionless distance is shown in Fig. 8.3. The predictions from Baumeister and Hamill and the solution by the method of relativistic transformation are within 12 percent of each other on the average. Close to the wave front, the error in the Chebyshev economization is expected to be small and verified accordingly. Close to the surface, the numerical error involved in the Chebyshev economization can be expected to be significant. This can be seen in Fig. 8.4 close to the surface. The method of relativistic transformation yields bounded solutions without any singularities. The transformation variable ψ is symmetric with respect to space and time. It transforms the PDE that governs the wave temperature into a Bessel differential equation. Three regimes are indentified in the solution—an inertial zero transfer regime, a regime characterized by Bessel composite function of zeroth order and first kind in space and time, and a regime characterized by modified Bessel composite function of zeroth order and first kind in space and time.

Earlier attempts by other investigators in order to obtain an analytical solution for the damped wave conduction and relaxation equation in an infinite cylindrical medium were made by using the method of Laplace transformation. Singularities were found in the results for a step change in temperature at the surface. In this study, the method of relativistic transformation is used in order to obtain analytical solution to infinite cylindrical coordinates for the case of a step change in boundary temperature. The transformation, $\eta = \tau^2 - X^2$, was found to transform the governing equation in wave temperature into a Bessel differential equation in one variable, i.e., the transformation variable. This was done for the case of infinite spherical medium as well. The governing equation for wave temperature from the governing equation for transient temperature can be obtained either by multiplying the transient temperature equation with $\exp(\tau/2)$ or by removing the damping component from the governing equation by a $u = w \exp(-\tau/2)$

substitution. The analytical solution for an infinite cylinder was characterized by a modified Bessel composite function in space and time of the first kind and half-order in the open interval of $\tau > X$. This is when the wave speed ($\sim r/t$) is smaller than the diffusion speed $\sqrt{\alpha}/\tau_r$. For values of times less than the dimensionless distance X, the solution is characterized by a Bessel composite function in space and time of the first kind and half-order. This is when the wave speed is greater than the diffusion speed. The inertial time lagging regime marked the third regime of transfer. For the infinite sphere, the solutions were characterized by a modified Bessel composite function in space and time of the first kind and first order and by a Bessel composite function in space and time of the first kind and first order for the open intervals of $\tau > X$ and $X > \tau$. The initial condition can be verified in the asymptotic limits of zero time. The transformation variable is symmetric with respect to space and time. No singularities were found in the analytical solutions for semi-infinite slab, infinite cylinder, and infinite sphere.

8.4   Finite Slab and Taitel Paradox

Taitel[10] considered a finite slab (Fig. 8.8) with two boundaries of width $2a$ heated from both sides. Both sides are maintained at a constant temperature T_s for times $t > 0$. At initial time $t = 0$, the temperature at all points in the slab is T_0. The governing equation is given by Eq. (8.2). The four conditions used by Taitel, two in space and two in time, that are needed to completely describe a hyperbolic PDE, i.e., second order with respect to space and with respect to time, were

$$t = 0 \quad -a < x < +a \quad T = T_0 \quad \text{or } u = 1 \tag{8.82}$$

$$t > 0 \quad x = \pm a \quad T = T_s \quad u = 0 \tag{8.83}$$

$$t = 0 \quad \partial u/\partial \tau = 0 \tag{8.84}$$

Taitel solved for Eq. (8.2) and for the conditions stated above obtained the analytical solution for damped wave conduction and

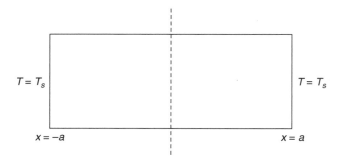

$T = T_s$ $T = T_s$

$x = -a$ $x = a$

FIGURE 8.8   Finite slab with two boundaries heated from both sides.

relaxation in a finite slab. The solution obtained by Taitel, for the centerline temperature of the finite slab, is given in the following. He considered a constant wall temperature and the initial time conditions included a $\partial T/\partial t = 0$ term in addition to the initial temperature condition. The solution he presented is as follows:

$$u = \sum_0^\infty b_n \exp(-\tau/2)\exp(-\tau/2\sqrt{(1-4(2n+1)^2\pi^2\alpha\tau_r/a^2)})$$
$$+ \sum_0^\infty c_n \exp(-\tau/2)\exp(+\tau/2\sqrt{(1-4(2n+1)^2\pi^2\alpha\tau_r/a^2)}) \qquad (8.85)$$

Multiplying both sides of =Eq. (8.85) by $\exp(\tau/2)$

$$u\,\exp(\tau/2) = W = \sum_0^\infty b_n \exp(-\tau/2\sqrt{(1-4(2n+1)^2\pi^2\alpha\tau_r/a^2)})$$
$$+ \sum_0^\infty c_n \exp(\exp(+\tau/2\sqrt{(1-4(2n+1)^2\pi^2\alpha\tau_r/a^2)})) \qquad (8.86)$$

At infinite times, the LHS of Eq. (8.86) is 0 times ∞ and is zero. The RHS does not vanish. For certain values of space and time, Taitel found that the analytical solution predicted values of temperature above the surface temperature. This is referred to in the literature as the temperature overshoot paradox. The temperature overshoot may be because of the growing exponential term in the previous expression.

8.4.1 Final Condition in Time for a Finite Slab

Consider the finite slab shown in Fig. 8.9 subject to the following four conditions: two in space and two in time that are required to complete a problem in hyperbolic PDE that is second order with respect to space and second order with respect to time.

$$t = 0 \quad -a < x < +a \quad T = T_0 \quad \text{or } u = 1 \qquad (8.87)$$
$$t > 0 \quad x = \pm a \quad T = T_s \quad u = 0 \qquad (8.88)$$
$$t = \infty \quad u = 0 \qquad (8.89)$$

Equation (8.89) is the final condition in time. Equation (8.2) is now solved for as follows.

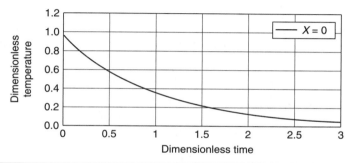

Figure 8.9 Centerline temperature in a finite slab at constant wall temperature (large a) ($a = 0.86\ m$, $\alpha = 10^{-5}\ m^2/s$, $\tau_r = 15\ s$).

Multiplying throughout Eq. (8.2) by $\exp(n\tau)$

$$\frac{\partial^2(ue^{n\tau})}{\partial X^2} = e^{n\tau}\frac{\partial u}{\partial \tau} + e^{n\tau}\frac{\partial^2 u}{\partial \tau^2} \tag{8.90}$$

Let $w = ue^{n\tau}$. Then,

$$\frac{\partial w}{\partial \tau} = e^{n\tau}\frac{\partial u}{\partial \tau} + ne^{n\tau}u = nw + e^{n\tau}\frac{\partial u}{\partial \tau}$$

$$\frac{\partial^2 w}{\partial \tau^2} = n\frac{\partial w}{\partial \tau} + ne^{n\tau}\frac{\partial u}{\partial \tau} + e^{n\tau}\frac{\partial^2 u}{\partial \tau^2} \tag{8.91}$$

Combining Eqs. (8.91) and (8.90),

$$\frac{\partial^2 w}{\partial X^2} = \frac{\partial w}{\partial \tau} - nw + \frac{\partial^2 w}{\partial \tau^2} - 2n\frac{\partial w}{\partial \tau} + n^2w \tag{8.92}$$

For $n = 1/2$, Eq. (8.90) becomes

$$\frac{\partial^2 w}{\partial X^2} = \frac{\partial^2 w}{\partial \tau^2} - \frac{w}{4} \tag{8.93}$$

w in Eq. (8.93) is the wave temperature. Equation (8.93) can be solved by the method of separation of variables. Let

$$u = V(\tau)\,\phi(X) \tag{8.94}$$

Equation (8.93) becomes

$$\phi''(X)/\phi(X) = (V'(\tau) + V''(\tau))/V(\tau) = -\lambda_n^2 \tag{8.95}$$

$$\phi(X) = c_1\sin(\lambda_n X) + c_2\cos(\lambda_n X) \tag{8.96}$$

From the boundary conditions,

$$\text{At } X = 0, \partial\phi/\partial X = 0, \quad \text{So, } c_1 = 0 \tag{8.97}$$

$$\phi(X) = c_1\text{Cos}(\lambda_n X) \tag{8.98}$$

$$0 = c_1\text{Cos}(\lambda_n X_a) \tag{8.99}$$

$$(2n-1)\pi/2 = \lambda_n X_a \tag{8.100}$$

$$\lambda_n = (2n-1)\pi\sqrt{(\alpha\tau_r)}/2a, n = 1,2,3,\ldots \tag{8.101}$$

The time domain solution would be

$$V = \exp(-\tau/2)\left(c_3\exp\left(\sqrt{\left(1/4 - \lambda_n^2\right)^\tau}\right)\right.$$

$$\left. + c_4\exp\left(-\sqrt{\left(1/4 - \lambda_n^2\right)^\tau}\right)\right) \tag{8.102}$$

or

$$Vexp(\tau/2) = \left(c_3 \exp\left(\sqrt{(1/4 - \lambda_n^2)}\tau\right)\right.$$
$$\left. + c_4 \exp\left(-\sqrt{(1/4 - \lambda_n^2)}\tau\right)\right) \tag{8.103}$$

From the final condition, $u = 0$ at infinite time. Therefore, $V\phi \exp(\tau/2) = W$, the wave temperature at infinite time. The wave temperature is that portion of the solution that remains after dividing the damping component from either the solution or the governing equation. For any nonzero ϕ, it can be seen that at infinite time the LHS of Eq. (8.103) is a product of zero and infinity and a function of x and is zero. Hence, the RHS of Eq. (8.103) is also zero and hence in Eq. (8.102) c_3 needs to be set to zero. Hence,

$$u = \sum_1^\infty c_n \exp(-\tau/2)\exp\left(\sqrt{(1/4 - \lambda_n^2)}\tau\right)\cos(\lambda_n X) \tag{8.104}$$

where λ_n is described by Eq. (8.101), C_n can be shown using the orthogonality property to be $4(-1)^{n+1}/(2n - 1)\pi$. It can be seen that Eq. (8.105) is bifurcated. As the value of the thickness of the slab changes, the characteristic nature of the solution changes from monotonic exponential decay to subcritical damped oscillatory. For $a < \pi\sqrt{(\alpha\tau_r)}$, even for $n = 1$, $\lambda_n > 1/2$. This is when the argument within the square root sign in the exponentiated time domain expression becomes negative and the result becomes imaginary. Using De Moivre's theorem and taking real part for small width of the slab,

$$u = \sum_1^\infty c_n \exp(-\tau/2)\cos\left(\sqrt{(\lambda_n^2 - 1/4)}\tau\right)\cos(\lambda_n X) \tag{8.105}$$

Equations (8.104) and (8.105) can be seen to be well bounded. Equations (8.104) and (8.105) become zero at long times. This would be time taken to reach steady state. Thus, for $a \geq \pi\sqrt{(\alpha\tau_r)}$

$$u = \sum_1^\infty c_n \exp(-\tau/2)\exp\left(-\sqrt{(1/4 - \lambda_n^2)}\tau\right)\cos(\lambda_n X) \tag{8.106}$$

where $c_n = 4(-1)^{n+1}/(2n - 1)\pi$ and $\lambda_n = (2n - 1)\pi\sqrt{(\alpha\tau_r)}/2a$

The centerline temperature for a particular example is shown in Fig. 8.10. Eight terms in the infinite series given in Eq. (8.106) were taken and the values calculated on a 1.9 GHz Pentium IV desktop personal computer. The number of terms was decided on the incremental change or improvement obtained by doubling the number of terms. The number of terms was arrived at a 4 percent change in the dimensionless temperature.

The Taitel paradox is obviated by examining the final steady state condition and expressing the state in mathematical terms. The W term, which is the dimensionless temperature upon removal of the damping term, needs to go to zero at infinite time. This resulted in

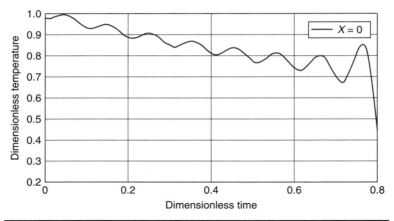

FIGURE 8.10 Centerline temperature in a finite slab at constant wall temperature (CWT) for small a ($a = 0.001$ m, $\alpha = 10^{-5}$ m^2/s; $\tau_r = 15$ s).

our solution, which is different from previous reports and is well bounded. The use of FINAL condition may be what is needed for this problem to be used extensively in engineering analysis without being branded as violating the second law of thermodynamics. The conditions that were touted as violations of the second law are not physically realistic. A bifurcated solution results. For small width of the slab, $a < \pi\sqrt{(\alpha\tau_r)}$, the transient temperature is subcritical damped oscillatory. The centerline temperature is shown in Fig. 8.11.

$$u = \sum_{0}^{\infty} c_n \exp(-\tau/2)\cos\left(\sqrt{(\lambda_n{}^2 - 1/4)}\tau\right)\cos(\lambda_n X) \qquad (8.107)$$

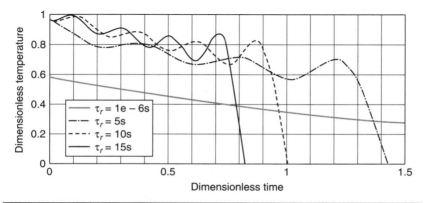

FIGURE 8.11 Centerline temperature in a finite slab at CWT for large relaxation times ($a = 0.001$ m, $\alpha = 10^{-5}$ m^2/s).

The subcritical damped oscillations in the centerline temperature at various values of large relaxation times are shown in Fig. 8.11. The relaxation time value greater than which the subcritical damped oscillations can be seen is given by

$$\tau_r > \frac{a^2}{\pi^2 \alpha} \tag{8.108}$$

8.4.2 Finite Sphere Subject to Constant Wall Temperature

Consider a sphere at initial temperature T_0. The surface of the sphere is maintained at a constant temperature T_s for times greater than 0. The heat propagative velocity is given as the square root of the ratio of thermal diffusivity and relaxation time,

$$V_h = \sqrt{\frac{\alpha}{\tau_r}}$$

The initial, final, and boundary conditions are

$$t = 0 \quad 0 \le r < R \quad T = T_0 \tag{8.109}$$
$$t = \infty \quad 0 \le r < R \quad T = T_s \tag{8.110}$$
$$t > 0 \quad r = 0 \quad \partial T / \partial r = 0 \tag{8.111}$$
$$t > 0 \quad r = R \quad T = T_s \tag{8.112}$$

The governing equation can be obtained by eliminating q_r between the generalized Fourier's law of heat conduction and the equation from energy balance of in-out accumulation. This is achieved by differentiating the constitutive equation with respect to r and the energy equation with respect to t and eliminating the second cross derivative of q with respect to r and time. Thus,

$$\tau_r \frac{\partial^2 T}{\partial t^2} + \frac{\partial T}{\partial t} = \alpha \frac{\partial^2 T}{\partial r^2} + \frac{2}{r} \frac{\partial T}{\partial r} \tag{8.113}$$

In the dimensionless form, Eq. (34) can be written in cylindrical coordinates as

$$\frac{\partial u}{\partial \tau} + \frac{\partial^2 u}{\partial \tau^2} = \frac{\partial^2 u}{\partial X^2} + \frac{2}{X} \frac{\partial u}{\partial X} \tag{8.114}$$

The solution is obtained by the method of separation of variables. First the damping term is removed by the substitution, $u = e^{-\tau/2}w$. By this substitution, Eq. (8.113) becomes

$$-\frac{w}{4} + \frac{\partial^2 w}{\partial \tau^2} = \frac{\partial^2 w}{\partial X^2} + \frac{2}{X} \frac{\partial w}{\partial X} \tag{8.115}$$

The method of separation of variables can be used to obtain the solution of Eq. (8.115).

$$\text{Let } w = V(\tau)\, \phi\, (X) \tag{8.116}$$

Plugging Eq. (8.116) into Eq. (8.115), and separating the variables that are a function of X only and τ only, the following two ordinary differential equations, one in space and another in time, are obtained:

$$\frac{d^2\phi}{dX^2} + \frac{2}{X}\frac{d\phi}{dX} + \lambda^2\phi = 0 \tag{8.117}$$

$$\frac{d^2V}{d\tau^2} = \left(\frac{1}{4} - \lambda^2\right) = 0 \tag{8.118}$$

The solution for Eq. (8.117) is the Bessel function of half-order and first kind:

$$\phi = c_1 J_{1/2}\,(\lambda\, X) + c_2 J_{-1/2}\,(\lambda\, X) \tag{8.119}$$

It can be seen that $c_2 = 0$ as the concentration is finite at $X = 0$. Now from the boundary condition (BC) at the surface,

$$\phi = c_1 J_{1/2}\left(\frac{\lambda R}{\sqrt{\alpha\tau_r}}\right) + c_2 J_{-1/2}\left(\frac{\lambda R}{\sqrt{\alpha\tau_r}}\right) \tag{8.120}$$

$$\frac{\lambda_n R}{\sqrt{\alpha\tau_r}} = (n-1)\pi \ \text{ for } n = 1, 2, 3 \ldots \tag{8.121}$$

The solution for Eq. (8.120) is the sum of two exponentials in time, one that decays with time and another that grows exponentially with time.

$$V = c_3 \exp\left(\tau\sqrt{0.25 - \lambda_n^2}\right) + c_4 \exp\left(-\tau\sqrt{0.25 - \lambda_n^2}\right) \tag{8.122}$$

The term containing the positive exponential power exponent will drop out because with increasing time, the system may be assumed to reach steady state and the points within the sphere will always have temperature values less than that at the boundary. From the final condition in time, i.e., at steady state,

$$w = u e^{\tau/2} \tag{8.123}$$

Thus, w will have to be zero at infinite time. Thus, c_3 in Eq. (8.15) is found to be zero. The term containing the positive exponential power exponent will drop out, as with increasing time the system may be assumed to reach steady state and the points within the sphere will always have temperature values less than that at the boundary.

Thus,

$$V = c_4 \exp\left(-\tau\sqrt{0.25 - \lambda_n^2}\right) \tag{8.124}$$

or

$$u = \sum_0^\infty c_n J_{1/2}(\lambda_n X) \exp\left(\frac{-\tau}{2} - \tau\sqrt{0.25 - \lambda_n^2}\right) \tag{8.125}$$

The c_n can be solved for from the initial condition by using the principle of orthogonality for Bessel functions. At time zero, the LHS and RHS are multiplied by $J_{1/2}(\lambda_m X)$. Integration between the limits of 0 and R is performed. When n is not m, the integral is zero from the principle of orthogonality. Thus, when $n = m$,

$$c_n = \frac{-\int_0^R J_{1/2}(\lambda_n X)}{\int_0^R J_{1/2}^2(c\lambda_n X)} \tag{8.126}$$

It can be noted from Eq. (8.45) that when

$$1/4 < \lambda_n^2 \tag{8.127}$$

the solution will be periodic with respect to the time domain. This can be obtained by using De Moivre's theorem and obtaining the real part to $\exp(-i\tau\sqrt{\lambda_n^2 - 0.25})$) Thus, for materials with relaxation times greater than a certain limiting value, the solution for temperature will exhibit subcritical damped oscillations. Thus,

$$\tau_r > \frac{R^2}{12.57\alpha} \tag{8.128}$$

Thus, a bifurcated solution is obtained. Also from Eq. (8.46) it can be seen that all terms in the infinite series will be periodic, i.e., even for $n = 2$ when Eq. (8.48) is valid.

$$u = \sum_0^\infty c_n J_{1/2}(\lambda_n X) \cos\left(\tau\sqrt{\lambda_n^2 - 0.25}\right) \tag{8.129}$$

Thus, the transient temperature profile in a sphere is obtained for a step change in temperature at the surface of the sphere using the modified Fourier's heat conduction law. For materials with relaxation times greater than $R^2/12.57\alpha$, subcritical damped oscillations can be seen in the transient temperature profile. The exact solution for transient temperature profile using finite speed heat conduction is derived by the method of separation of variables. It is a bifurcated solution. For certain values of λ the time portion of the solution is cosinous and damped and for others it is an infinite series of Bessel functions of the first kind and half-order and decaying exponential in time. In addition, it can be shown that for terms in the infinite series with n greater than 2, the contribution to the solution will be periodic for small R. The exact solution is bifurcated.

8.4.3   Finite Cylinder Subject to Constant Wall Temperature

Consider a cylinder at initial temperature T_0. The surface of the sphere is maintained at a constant temperature T_s for times greater than 0. The heat propagative velocity is given as the square root of the ratio of thermal diffusivity and relaxation time,

$$V_h = \sqrt{\frac{\alpha}{\tau_r}}$$

The initial, final, and boundary conditions are the same as given for the sphere. The governing equation can be obtained by eliminating q_r between the generalized Fourier's law of heat conduction and the equation from energy balance of in-out accumulation. This is achieved by differentiating the constitutive equation with respect to r and the energy equation with respect to t and eliminating the second cross derivative of q with respect to r and time. Thus,

$$\tau_r \frac{\partial^2 T}{\partial t^2} + \frac{\partial T}{\partial t} = \alpha \frac{\partial^2 T}{\partial r^2} + \frac{\alpha}{r} \frac{\partial T}{\partial r} \tag{8.130}$$

The governing equation in the dimensionless form is then

$$\frac{\partial u}{\partial \tau} + \frac{\partial^2 u}{\partial \tau^2} = \frac{\partial^2 u}{\partial X^2} + \frac{1}{X} \frac{\partial u}{\partial X} \tag{8.131}$$

The solution is obtained by the method of separation of variables. First the damping term is removed by the substitution, $u = e^{-\tau/2}w$.

$$-\frac{w}{4} + \frac{\partial^2 w}{\partial \tau^2} = \frac{\partial^2 w}{\partial X^2} + \frac{1}{X} \frac{\partial w}{\partial X} \tag{8.132}$$

The method of separation of variables can be used to obtain the solution of Eq. (8.52). Let

$$w = V(\tau)\,\phi\,(X) \tag{8.133}$$

Plugging Eq. (8.54) into Eq. (8.53), and separating the variables that are a function of X only and τ only, the following two ordinary differential equations, one in space and another in time, are obtained

$$\frac{d^2\phi}{dX^2} + \frac{1}{X}\frac{d\phi}{dX} + \lambda^2\phi = 0 \tag{8.134}$$

$$\frac{d^2V}{d\tau^2} = \left(\frac{1}{4} - \lambda^2\right) = 0 \tag{8.135}$$

The solution to Eq. (8.55) can be seen as a Bessel function of the zeroth order and first kind and Bessel function of the zeroth order and second kind.

$$\phi = c_1 J_0\left(\frac{\lambda R}{\sqrt{\alpha\tau_r}}\right) + c_2 Y_0\left(\frac{\lambda R}{\sqrt{\alpha\tau_r}}\right) \tag{8.136}$$

It can be seen that $c_2 = 0$ as the temperature is finite at $X = 0$. Now from the BC at the surface,

$$\frac{\lambda_n R}{\sqrt{\alpha\tau_r}} = 2.4048 + (n-1)\pi \text{ for } n = 1, 2, 3, \ldots \tag{8.137}$$

The solution for Eq. (8.56) is the sum of two exponentials in time, one that decays with time and another that grows exponentially with time.

$$V = c_3 \exp\left(\tau\sqrt{0.25 - \lambda_n^2}\right) + c_4 \exp\left(-\tau\sqrt{0.25 - \lambda_n^2}\right) \tag{8.138}$$

The term containing the positive exponential power exponent will drop out as with increasing time the system may be assumed to reach steady state and the points within the sphere will always have temperature values less than that at the boundary. From the final condition in time, i.e., at steady state,

$$w = ue^{\tau/2} \tag{8.139}$$

Thus, w will have to be zero at infinite time. Thus, c_3 in Eq. (8.59) is found to be zero. The term containing the positive exponential power exponent will drop out, as with increasing time the system may be assumed to reach steady state and the points within the sphere will always have temperature values less than that at the boundary. Thus,

$$V = c_4 \exp\left(-\tau\sqrt{0.25 - \lambda_n^2}\right) \tag{8.140}$$

or

$$u = \sum_0^\infty c_n J_0(\lambda_n X) \exp\left(\frac{-\tau}{2} - \tau\sqrt{0.25 - \lambda_n^2}\right) \qquad (8.141)$$

The c_n can be solved for from the initial condition by using the principle of orthogonality for Bessel functions. At time zero, the LHS and RHS are multiplied by $J_0(\lambda_m X)$. Integration between the limits of 0 and R is performed. When n is not m, the integral is zero from the principle of orthogonality. Thus, when $n = m$,

$$c_n = \frac{-\int_0^R J_0(\lambda_n X)}{\int_0^R J_0^2(c\lambda_n X)} \qquad (8.142)$$

It can be noted from Eq. (8.34) that when

$$1/4 < \lambda_n^2 \qquad (8.143)$$

the solution will be periodic with respect to time domain. This can be obtained by using De Moivre's theorem and obtaining the real part to $\exp(-i\tau\sqrt{\lambda_n^2 - 0.25}\,)$. Thus, for materials with relaxation times greater than a certain limiting value, the solution for temperature will exhibit subcritical damped oscillations.

$$\tau_r > \frac{R^2}{9.62\alpha} \qquad (8.144)$$

Thus, a bifurcated solution is obtained. Also from Eq. (8.62) it can be seen that all terms in the infinite series will be periodic, i.e., even for $n = 2$ when Eq. (8.64) is valid,

$$u = \sum_0^\infty c_n J_0(\lambda_n X) \cos\left(\tau\sqrt{\lambda_n^2 - 0.25}\right) \qquad (8.145)$$

Thus, the transient temperature profile in a cylinder is obtained for a step change in temperature at the surface of the cylinder using the modified Fourier's heat conduction law. For materials with relaxation times greater than $(R^2/9.62\alpha)$, where R is the radius of the cylinder, subcritical damped oscillations can be seen in the transient temperature profile. The exact solution for a finite cylinder subject to constant wall temperature using finite speed heat conduction is derived by the method of separation of variables. It is a bifurcated solution. For certain values of λ, the time portion of the solution is cosinous and damped and for others it is an infinite series of Bessel functions of the first kind and half-order and decaying exponential in time. In addition, it can be shown that for terms in the infinite series with n greater than 2, the contribution to the solution will be periodic for small R.

The temperature overshoot found in the analytical solution of Taitel for the case of a finite slab subject to constant wall temperature was the cause for alarm for a possible violation of the second law of thermodynamics. In this study, the final condition in time is posed as one of the two space conditions and two time conditions needed in order to fully describe a second order hyperbolic partial differential equation in two variables. In addition to the initial time condition, the constraint from the steady state attainment is translated to a fourth time condition. The wave dimensionless temperature has to become zero at steady state and the wave temperature itself has to attain equilibrium temperature. When this condition is applied, a growing exponential in time vanishes and a well-bounded solution results for a finite sphere and a finite cylinder. Taitel used a condition at time zero that the time derivative of the temperature will be zero. This includes any initial temperature distribution. It turns out that this cannot be a physically realistic fourth condition. The fourth condition in this study comes from what can be expected at steady state. The time derivative of temperature at zero time may have to be calculated from the model solution. In terms of degrees of freedom in time conditions, the constraint from steady state has to take precedence. The method of separation of variables was used to obtain the analytical solution. The solutions were found to be bifurcated for all three cases of finite slab, finite sphere, and finite cylinder.

When the relaxation time of the material under consideration becomes large, the temperature can be expected to undergo oscillations in time domain. These oscillations were found to be subcritical damped oscillatory. For a finite sphere, when the relaxation times are greater than $R^2/(12.57\alpha)$, the solution becomes subcritical damped oscillatory from monotonic exponential decay in time and is given as an infinite Bessel series solution of the half-order and first kind. For a finite cylinder, when the relaxation times are greater than $R^2/(9.62\alpha)$, the solution becomes subcritical damped oscillatory from monotonic exponential decay in time and given as an infinite Bessel series solution of the zeroth order and first kind. For a finite slab, when the relaxation times are greater than $a^2/\pi\alpha$, the solution becomes subcritical damped oscillatory from monotonic exponential decay in time and is given by an infinite Fourier series solution.

The expressions for heat flux can be obtained from the energy balance equation and the convergence of the infinite series confirmed at the surface. Thus, the singularities found in the solution to the Fourier parabolic equations for the same geometry are now absent in the solution to the damped wave conduction and relaxation hyperbolic equations. The main conclusions from the study are as follows:

1. Use of final condition in time leads to bounded solutions.

2. The temperature overshoot problem can be attributed to use of a physically unrealistic time condition.

3. Analytical solution obtained for finite sphere, finite cylinder, and finite slab is found to be bifurcated.

4. For materials with large values of relaxation times, such as given in Eq. (8.108) for a finite slab, Eq. (8.128) for the case of a finite sphere, and Eq. (8.144) for the case of finite cylinder, subcritical damped oscillations in temperature, can be found.

8.5 Summary

Nanoscale effects in time domain are important in a number of applications. Recent experimental studies have been reported to capture the conformational changes of proteins at the nanosecond time scale using a synchrotron based TR-WAXS instrument. Fourier's law of heat conduction, Fick's law of mass diffusion, Newton's law of viscosity, and Ohm's law of electricity are laws derived from empirical observations at steady state. There are six reasons to seek a generalized Fourier's law of heat conduction:

1. The microscopic theory of reversibility of Onsager is violated

2. Singularities were found in a number of important industrial applications of the transient representation of temperature, concentration, and velocity

3. Development of Fourier's law was from observations at steady state

4. Overprediction of theory to experiment has been found in a number of industrial applications

5. Landau and Lifshitz observed the contradiction of the infinite speed of propagation of heat with Einstein's light speed barrier

6. Fourier's law breaks down at the Casimir limit. The generalized Fourier's law of heat conduction is given by Eq. (8.1) and was postulated independently by Cattaneo and Vernotte.

Consider a semi-infinite medium at an initial temperature of T_0 subject to a constant surface temperature boundary condition for times greater than 0. The hyperbolic PDE that forms the governing equation of heat conduction is solved for a new method called relativistic transformation of coordinates. The hyperbolic PDE is multiplied by $e^{\tau/2}$ and transformed into another PDE in wave temperature. This PDE is converted to an ODE by the transformation variable that is spatio-temporal symmetric. The resulting ODE is seen to be a generalized Bessel differential equation. The solution from this approach is within 12 percent of the exact solution obtained by Baumeister and Hamill using the method of Laplace transforms. There are no singularities in the solution. There are three regimes to the solution: (1) inertial regime, (2) regime characterized by Bessel composite function of the zeroth order and first kind, and (3) a regime characterized by modified Bessel composite function of the zeroth order and first kind. Expressions for penetration length and inertial lag time are developed. The comparison between the solution from the method of relativistic transformation of coordinates and the method of Laplace transforms was made by use of Chebyshev polynomial approximation and numerical integration. The dimensionless temperature as a function of dimensionless distance for the parabolic, hyperbolic solved by relativistic transformation and hyberbolic solved by Laplace transforms is shown in Fig. 8.3.

In a similar manner, the exact solution to the hyperbolic PDE is solved for by the method of relativistic transformation of coordinates for the infinite cylindrical and infinite spherical medium.

For the case of heating a finite slab, the Taitel paradox problem is revisited. Taitel found that when the hyperbolic PDE was solved for the interior, temperature in the slab was found to exceed the wall temperature of the slab. This is in violation of the second law of thermodynamics. By use of the final condition in time at steady state, the wave temperature was found to be become zero at steady state. This condition when mathematically posed as the fourth condition for the second order PDE leads to well-bounded solutions within the bounds of the second law of thermodynamics. For systems with large relaxation times, i.e.,

$$\tau_r > \frac{a^2}{\pi^2 \alpha}$$

subcritical damped oscillations can be seen in the temperature. This is shown in Fig. 8.11. In a similar manner, the transient temperatures for a finite sphere and finite cylinder are also derived.

Review Questions

1. Distinguish between the wave and Fourier regimes.

2. Examine the problem of heating an infinite medium with constant thermal diffusivity from a cylindrical surface with a radius R. Assume a dimensionless heat flux at the wall as 1. Obtain the transient temperature using the parabolic Fourier equation. Is there a singularity in the solution expression?

3. Examine the problem of heating an infinite medium with constant thermal diffusivity from a spherical surface with a radius R. Assume a dimensionless heat flux at the wall as 1. Obtain the transient temperature using the parabolic Fourier equation. Is there a singularity in the solution expression?

4. Would Nernst's observation of thermal inertial and oscillatory discharge be a seventh reason for seeking a generalized Fourier's law of heat conduction?

5. Can the generalized Fourier's law of heat conduction be derived from kinetic theory of gases? If so, what is the physical significance of the ballistic term?

6. Can the generalized Fourier's law of heat conduction be derived from Stoke's Einstein expression[11] for diffusion coefficients? If so, what is the physical significance of the ballistic term?

7. Can the generalized Fourier's law of heat conduction be derived from the free electron theory? If so, what is the physical significance of the ballistic term?

8. What is the time taken to reach the steady state in the problem of heating a finite slab when solved for by the hyperbolic PDE? How is this different from the solution from the parabolic PDE?

9. How would the solution to Question 8 change when the boundary condition of the finite slab is changed from constant wall temperature to convective boundary condition?

10. Can the method of relativistic transformation of coordinates be used to solve for the transient temperature in a semi-infinite medium subject to constant wall temperature in Cartesian coordinates in three dimensions? If so, what would the transformation variable be?

11. Can the method of relativistic transformation of coordinates be used to solve for the transient temperature in an infinite medium subject to constant wall temperature in cylindrical coordinates in three dimensions? If so, what would the transformation variable be?

12. Can the method of relativistic transformation of coordinates be used to solve for the transient temperature in an infinite medium subject to constant wall temperature in spherical coordinates in three dimensions? If so, what would the transformation variable be?

13. What are the expressions for penetration distance and inertial lag times for the transient temperature in a semi-infinite medium subject to constant wall temperature in one-dimensional Cartesian coordinates?

14. What are the expressions for penetration distance and inertial lag times for the transient temperature in an infinite medium subject to constant wall temperature in one-dimensional cylindrical coordinates?

15. What are the expressions for penetration distance and inertial lag times for the transient temperature in an infinite medium subject to constant wall temperature in one-dimensional spherical coordinates?

16. What are the expressions for penetration distance and inertial lag times for the transient temperature in a semi-infinite medium subject to constant wall temperature in three-dimensional Cartesian coordinates?

17. What are the expressions for penetration distance and inertial lag times for the transient temperature in an infinite medium subject to constant wall temperature in three-dimensional cylindrical coordinates?

18. What are the expressions for penetration distance and inertial lag times for the transient temperature in an infinite medium subject to constant wall temperature in three-dimensional spherical coordinates?

19. What is the time taken to reach the steady state in the problem of heating a finite cylinder when solved for by the hyperbolic PDE? How is this different from the solution from the parabolic PDE?

20. What would the solution to Question 19 be should the boundary condition of the finite cylinder be changed from constant temperature to convective boundary condition?

21. What is the time taken to reach the steady state in the problem of heating a finite sphere when solved for by the hyperbolic PDE? How is this different from the solution from the parabolic PDE?

22. What would the solution to Question 21 be should the boundary condition of the finite sphere be changed from constant temperature to convective boundary condition?

23. In the analysis of transient temperature using the generalized Fourier's law of heat conduction in one-dimensional Cartesian coordinates in a semi-infinite medium, say that the temperature in the interior point p is given as a function of time. Obtain the general solution for the transient temperature. What is the temperature at $X = 0$?

24. Obtain the transient temperature in a right circular cone of infinite height with a constant apex temperature for times greater than 0. What is the effect of the change in area with distance from apex of the cone?

25. Consider a finite slab subject to the convective boundary condition. Using a space averaged expression for temperature, obtain the governing equation for transient temperature for the slab using the generalized Fourier's law of heat conduction equation. The heat transfer coefficient is periodic in time and expressed as:

$$h = h_0 + h_A \cos(\omega t)$$

Derive the transient temperature for the entire slab. Comment on the nature of the solution for materials with large relaxation times. Discuss the attenuation and phase lag.

26. Consider the Earth's crust heated by the sun. The initial temperature of the earth is at T_0 imposed by a periodic temperature at the crust by $T_0 + T_s\cos(\omega t)$.

27. By the method of Laplace transforms, obtain the transient temperature in an infinite cylindrical medium subject to constant wall temperature for times greater than 0.

28. By the method of Laplace transforms, obtain the transient temperature in an infinite spherical medium subject to constant wall temperature for times greater than 0.

29. How does the solution by the method of relativistic transformation compare with the solution obtained in Question 27?

30. How does the solution by the method of relativistic transformation compare with the solution in Question 28?

31. At what values of relaxation times of the materials above which subcritical damped oscillations in temperature would be expected for a finite cylinder subject to constant wall temperature?

32. At what values of relaxation times of the materials above which subcritical damped oscillations in temperature would be expected for a finite sphere subject to constant wall temperature?

33. Repeat Question 31 for the convective boundary condition.

34. Repeat Question 32 for the convective boundary condition.

35. What happens to the convex temperature profile obtained from Fourier equation when hyperbolic PDE is used? What is the physical significance of the change from concave to convex curvature in the transient temperature?

36. What happens at the wave front in a semi-infinite medium in one-dimensional Cartesian coordinates subject to constant wall temperature boundary conditions?

37. What happens at the wave front in a semi-infinite medium in one-dimensional Cartesian coordinates subject to constant wall flux boundary conditions?

38. What happens at the wave front in a semi-infinite medium in three-dimensional Cartesian coordinates subject to constant wall temperature boundary conditions?

39. What happens at the wave front in an infinite medium in three-dimensional cylindrical coordinates subject to constant wall temperature boundary conditions?

40. What happens at the wave front in an infinite medium in three-dimensional spherical coordinates subject to constant wall temperature boundary conditions?

References

1. C. Borman, Proteins caught in the act, *Chem. Eng. News*, 6, 39, 11, 2008.
2. L. Onsager, Reciprocal relations in reversible processes, *Phys. Rev.*, 37, 405–426, 1931.
3. L. Landau and E. M. Lifshitz, *Fluid Mechanics*, Pergamon, Oxford, UK, 1987.
4. K. R. Sharma, *Damped Wave Conduction and Relaxation*, Elsevier, Amsterdam, The Netherlands, 2005.
5. J. C. Maxwell, On the dynamical theory of gases, *Phil. Trans. of Royal Society of London*, 157, 49, 1867.
6. D. D. Joseph and L. Preziosi, Heat waves, *Rev. Modern Phys.*, 61, 1, 41–73, 1989.

7. K. J. Baumeister and T. D. Hamill, Hyperbolic heat conduction equation—a solution for the semi-infinite medium, *ASME J. Heat Transfer*, 93, 1, 126–128, 1971.

8. K. J. Baumeister and T. D. Hamill, Hyperbolic heat conduction—a solution for the semi-infinite body problem, *J. Heat Transfer*, 91, 543–548, 1969.

9. K. R. Sharma, Manifestation of acceleration during transient heat conduction, *J. Thermophys. Heat Transfer*, 20, 4, 799–808, 2006.

10. Y. Taitel, On the parabolic, hyperbolic and discrete formulation of the heat conduction equation, *Intl. J. Heat Mass Transfer*, 15, 2, 369–371, 1972.

11. K. R. Sharma, On the derivation of a damped wave transport equation from free electron theory, *231st ACS National Meeting*, March 26–30, Atlanta, GA, 2006.

CHAPTER 9

Characterization of Nanostructres

Learning Objectives

- Principle of operation of SAXS, resolution achievable and applications
- TEM sample preparation, resolution achievable, applications, and principle of operation
- Topography generation using SPM, applications
- QD analysis using microwave spectroscopy
- Raman microscope design, advantages
- STM evolution, applications
- Helium ion microscopes (HeIMs) and world record in resolution

9.1 Overview

As interest in nanostructuring operations increases, there is a need to characterize these structures and develop suitable instruments to study them. The resolution limits of optical microscopes would be of the order of the wavelength of light. Thus, per the Raleigh criterion, the resolution limit using optical devices is 200 nm, half the wavelength of light. Devices with x-ray sources are needed for characterizing nanostructures. In Table 9.1, the resolution size of different equipment is given. This chapter is devoted to the equipment needed to characterize nanostructures.

9.2 Small-Angle X-Ray Scattering (SAXS)

The nanoscale structure of particle systems can be studied using SAXS devices in terms of parameters such as average particle size, particle shape, particle size distribution, and surface-to-volume ratio.

10^{-14}	10^{-12}	10^{-10}	10^{-8}	10^{-6}	10^{-4}	10^{-2}	1
	Scanning probe microscope						
		Transmission electron microscope					
			Scanning electron microscope				
				Optical microscope			

TABLE 9.1   Resolution Limits of Different Microscopes

The x-ray source can be a synchrotron light, which provides high x-ray flux. Structural information in the scale of 5 to 25 nm can be characterized using SAXS. The test is nondestructive and crystalline samples are needed. A monochromatic source of x-rays is used to excite the sample and the scattered x-rays are detected by a two-dimensional flat x-ray detector. The structure is deduced from the scattering pattern. The weakly scattered intensity needs to be separated from the main beam with large strength. The difficulty increases with a decrease in desired angle. Divergent beams are produced from x-ray sources, which compound the problem. Beam focus is used to overcome the difficulties encountered. Collimation is relied upon, as beam focus is not readily accomplished. In point-collimation instruments, the x-ray beam is shaped by use of pinholes to a small spot with a circular or elliptical shape in the plane of detection. Scattered intensity is expressed as a function of scattering vector in the observed data. Porod's law states that the contribution to the scattering that comes from the interface between two phases and the intensity should drop with the fourth power of q if this interface is smooth: $I = kSq^{-4}$. The surface area S of the particles can be determined by SAXS. With a fractally rough surface area with a dimensionality d between 2 and 3, Porod's law becomes: $I = kS'q^{-(6-d)}$. The particle size distribution can also be obtained from the SAXS data. The Guinier approximation can be applied to the beginning of the scattering curve at small q values. The intensity in this regime would depend on the radius of gyration of the particle. The distance distribution function is calculated by use of Fourier transform. The distance distribution $p(r)$ is related to the frequency of certain distances r within the particle.

X-ray scattering has been used to study material structures that possess long-range order. The technique is based upon elastic scattering of x-rays by the structures under scrutiny. In order for SAXS to be an effective technique, deviations from average electron density have to be present in the sample. When these are concentrated in systems, the intensity in SAXS measurements is proportional to the Fourier transform of their form factor. The Guinier approximation can be used to show that the intensity variation with the scattering wave vector close to the origin is the same regardless of the shape and depends only on the size of the particle. An additional contribution from particle to particle correlation can be included in agglomerated systems and $I = F(q)S(q)$.

The structure factor is the measure of short-range order. A maximum is displayed at an angle reciprocal to the average of the first neighbor distance. SAXS use is also limited when thin-layered systems on thicker substrates are used. The deep penetration depth and greatly reduced signal-to-noise ratio are the limiting attributes of the instrument. When grazing incidence is used, the technique is called GISAXS. When grazing angle is varied, the penetration depth can be limited. This can provide information up to several 100 nm into the substrate, which was not accessible by other microscopy techniques.

SAXS was shown to be versatile[1] in characterizing nanostructures. It is a nondestructive technique. It can be used for one-dimensional structures such as thin films or thin multilayered systems or three-dimensional oriented nanoparticles of different chemical composition either buried in thin film or grown on a substrate surface. It is more advantageous over near field microscopy techniques. It is selected for material characterization in the 1 to 100 nm range.

SAXSess instrument from Anton-Paar can be used for structural characterization of nanomaterials. This model has been made faster by use of modern focusing multilayer optics reaching an intensity gain that is 20 times larger compared with previous models (Fig. 9.1).

FIGURE 9.1   SAXSess instrument from Anton-Paar.

The instrument features versatile sample holders for liquids, pastes, powders, polymers, and fibers, simultaneous measurements of up to 40°, and precise temperature control in the range of −150 to 300°C. Crystallinity and phase state can be determined using the instrument. Software is dedicated for fast results and software allows for data interpretation and model building. SAXSess can be used as a tool to study different samples such as colloidal liquids, protein solutions, and solid nanocomposites and polymer thin films. The micelle size, micelle shape, phase behavior, and inner structure of vesicle walls in surfactants can be examined using SAXSess. Surface-active agents are made up of molecules consisting of a hydrophobic part and a hydrophilic part. The molecule undergoes self-assembly and forms micelles and other nanostructures. These can be spherical, cylindrical, or lamellar in shape. Typical sample specimens studied are detergents, food additives and nutrients, pharmaceuticals, and personal care products.

In order to better understand the functioning of biological processes, information on the structure of proteins, enzymes, allergens, and biomembranes is needed. SAXSess has been chosen as a tool in biomaterials research to obtain structural information of proteins in solution, inner structure, aggregation state, and molecular weight. The instrument can be used to obtain shape or size distribution of dispersed particles, dispersion stability, particle nucleation, and aggregation state. Fiber structure can also be studied using SAXSess. The internal structure, crystallinity, specific surface, and orientation distribution, for instance, can be calculated using SAXSess. Catalyst properties such as specific surface, porosity, particle size, particle size distribution, and crystallinity have been studied using this instrument. Emulsion characteristics such as shape and inner structure, particle size distribution of droplets, emulsion stability at different temperatures, and transfer kinetics of encapsulated agents can be obtained. The shape and inner structure of polymers and nanocomposites, crystallinity, periodic nanostructures, and orientation can be looked at using this tool. The size distribution and shape of liquid crystals, the crystallinity, aggregate ordering, and orientation can be measured using the device.

SAXS is rapidly becoming a well-established analytical method for nanostructure analysis. When x-rays are allowed to penetrate the specimens, they are scattered at the interfaces of nanostructures. A scattering pattern is produced that is specific to the nanostructure. The method is nondestructive, less expensive, and requires less sample preparation. It allows for investigations of interactions between molecules in real time. These interactions may lead to self-assembly and large-scale structure changes on which material properties depend. Its applications have been found in a variety of fields such as emulsions, liquid crystals, and macromolecules to porous molecules and metal alloys. It is used in research, technology development, and statistical quality control.

Nanoscale particles scatter toward small angles. The SAXS pattern provides information on the overall size and shape of these particles. The orientation and nanostructure of the sample can be obtained. Atoms and interatomic distances scatter toward large angles. The obtained WAXS pattern provides information on the phase state, crystal symmetry, and the molecular structure.

True SWAXS is a feature of SAXSess that allows for simultaneous measurement continuously from small and wide angles up to 40° with a consistent high resolution and without changing the instrumental setup. The complete information on the nanostructure of the sample and its phase state such as crystalline or amorphous can be obtained in a single run. A high-intensity signal at the detector is produced. Nanostructures in the range of 0.2 to 150 nm can be investigated. The ease of handling of the facility aids the measurement procedure and provides for ready alignment. Line collimation is offered for rapid data acquisition of isotropic samples and point collimation is provided for studies of oriented samples in one device. Detection systems provide remarkable resolution. The device consists of the following components:

1. X-ray source—long-term stability and low operation costs
2. Advanced focusing multilayer optics
3. Enhanced block collimation system
4. Sample stage
5. Semitransparent beam stop
6. TrueSWAXS—single run feature
7. High-performance detector system
8. SAXSquant software system
9. Temperature control
10. Versatile sample holders

9.3   Transmission Electron Microscope (TEM)

TEM can resolve structures in the nanoscale region. Sample preparation for TEM studies is complex and requires other instruments. The sample specimens have to have thinness down to a few hundred nanometers depending on the operating voltage of the instrument. They must possess parallel surfaces. A thin section of 0.5 to 3.0 mm is cut from the bulk material using electric discharge machining or other such similar techniques and a rotating wire saw. The specimen is milled down to 50 μm thickness. Electropolishing and ion beam thinning are used to finish the sample to its final dimensions.

A high voltage of 100 to 300 kV is applied to a tungsten filament and an electron beam is produced. This is then accelerated down to the specimen. Electromagnetic coils are used to have the electron beam condensed. The electrons are allowed to pass through the specimen. As the electrons pass through the specimen, some are absorbed and some are scattered and change direction. Sample thinness is a critical parameter. With thick samples due to excessive absorption and diffraction, the transmission of electrons will not be permitted. Electron scattering is caused by differences in atomic crystal arrangement. The electron beam upon transmittal is focused with a magnetic lens and then amplified and projected on a fluorescent screen. An image is formed that is either a bright field image or a dark field image depending on whether the direct beam or the scattered beam is selected. Irregular atomic arrangement such as dislocations in material imperfections will appear as dark lines on the electron microscope screen.

High-resolution transmission electron microscope (HRTEM) can achieve resolution sizes down to 1 Å. Atomic level phenomena can be viewed using HRTEM techniques.

VG Microscopes has developed scanning transmission electron microscopy (STEM) instruments. The objective lens in these instruments is a strong electromagnetic lens. This allows for demagnification of the electron source and formation of a fine probe of the specimen. The excitation source is of the cold-field emission type for minimum size and maximum brightness. Noise is reduced by operation in vacuum of up to 10^{-10} torr. One hundred kilovolt source offers the energy for acceleration of the electrons. One to three condenser lenses precede the objective lens, as the application calls for the beam configuration and apertures may be selected. The fine probe is allowed to be scanned over the specimen. When the fine probe is passed through the specimen, a convergent beam diffraction pattern is formed on the distant plane of observation. This may be observed and recorded by using a suitable phosphor screen and charged couple device (CCD) cameras. The STEM signal is formed from a part of this pattern. This is then displayed on a cathode ray tube with a scan synchronized with that of the probe at the specimen level to form the magnified image. A bright field image is formed by detection of a portion of the central beam of the convergent beam electron diffraction (CBED) pattern. Dark field images may be obtained by detection of individual diffraction spots. Electrons scattered outside the central spot may be collected using annular detector.

The specimen is placed within the magnetic field of the objective lens for high-resolution STEM imaging. The focusing effect is produced by the magnetic field of the objective. A lens system is formed by use of two or more postspecimen lenses. One advantage of STEM instruments compared with TEM instruments is the design flexibility.

TEM has been used in life sciences and biomedical investigations because of its ability to view the finest cell cultures. It is used as a diagnostic tool in pathology laboratories of hospitals all over the world.

High-voltage/high-resolution TEMs manufactured by JEOL, Tokyo, Japan utilizing 200 keV to 1 MeV offer resolution sizes down to the size of an atom. This allows for imaging of atoms and design of materials with tailor-made properties. TEM can be used as an elemental analysis tool with the addition of energy dispersive x-ray analysis (EDXA) or energy loss spectrometry (ELS). Elements in areas less than 500 nm in diameter can be identified using this tool.

Figure 9.2 depicts a low resolution of a cryo-TEM image of surfactant peptides at pH 7 at a peptide concentration of 5 mg/mL. This is from a recent patent from MIT.[2] They discuss dipolar oligopeptides that self-assemble to form regular structures. Gold is localized upon the nanostructures that are formed from self-assembly. Different nanoarchitectures are formed from self-assembly. Short oligonucleotides with di- and triblock peptide copolymers with properties that mimic those found in surfactant molecules were synthesized. Well-defined structures of approximately 50 nm were prepared by self-assembly of short amphiphilic peptides. Such self-assembled structures can be modified by external parameters such as pH. Adequate design of the peptide allows fine-tuning the self-assembly properties and offers flexibility for various applications. One class of peptides has been designed and investigated for its ability to self-assemble spontaneously to form stable nanotubes. It consists of a hydrophobic and hydrophilic group. The lipophilic tail is made out

FIGURE 9.2
Cryo-TEM image of aqueous solution of V_6D (tetra valine aspartic acid).

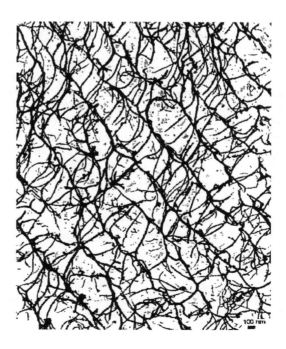

of alanine, valine, isoleucine, or leucine. The hydrophilic head is made out of charged amino acids such as lysine, arginine, histidine, aspartic acid, and glutamic acid. When dispersed in water, the amphiphilic peptides tend to self-assemble in the formation of a polar interface, which separates the hydrocarbon and water regions.

Cryo-TEM is performed at temperatures of −160 to −50°C. An electron beam is transmitted into the specimen and focused on it. The image contrasts are formed by scattering of electrons out of the beam and various magnetic lenses are allowed to perform in lieu of ordinary lenses in an optical microscope. As can be seen in Fig. 9.2, the diameter of the oligopeptide sequence is 10 to 15 nm. In order to obtain an aqueous solution of the oligopeptides, it is necessary to deprotonate the carboxylic groups by changing the pH with a solution of 0.1 N NaOH. Solubilization was found to begin at a pH of 5 to 6, depending on the amino acid sequence. It was found from cryo-TEM investigations that charged oligopeptides exist as a dense network of entangled nanotubes with diameter ranging from 25 to 50 nm. Cylindrical assemblers provide a three-dimensional transient network. Only two-dimensional projection of the peptide nanotubes were found to be imaged. Cylindrical morphologies were found for the V_6D oligopeptide structure (V = valine and D = aspartic acid). High axial ratios were found in the self-assemblies. The length extends to several microns. Many threefold junctions or branching connect the nanotubes, forming the final network. This kind of branching can be shown to be energetically unstable as shown in Fig. 9.2.

Branched supramolecular structures have been discussed in the literature. Branching has been found in aqueous surfactant solutions. Reverse structures such as lecithin organogels are also found to form threefold junctions. Patches are produced at branch points having mean curvature opposite to that of the portion far from the junction. Such branching points have been attributed as causative in the manifestation of viscoelastic properties of polymer-like systems. Branches vs. entanglements have been studied. Thus, tubular nanostructure can be constructed from self-assembly of oligopeptides. Fine-tuning of monomer properties will give rise to a wide range of nanostructures. These can result in development of novel biomaterials.

9.4   Scanning Electron Microscope

In 1935, using electron channeling contrast, Knoll obtained the world's first SEM image of silicon steel. It was first marketed in 1965 by Cambridge Instrument Company under the name Stereoscan. Magnification in SEM is over five orders of magnitude ranging from 25× to 250,000×. In SEMs, the image magnification is no longer a function of the objective lens. The electron beam is focused on a spot. As the electron gun is made to generate a beam with sufficiently smaller diameter, the need for condenser and objective lenses ceases. Similar to SPM, the magnification results from the ratio of the dimensions of the raster on the specimen and the raster on the display device. The current supplied to the scanning coils determines the magnification.

The spatial resolution of the SEM depends on the wavelength of electrons and the electro-optical system, which produces the scanning beam. The extent to which the material interacts with the electron beam is also an important consideration. Resolution of atom scale is not possible as can be in the case of TEM. However, the advantage of using SEM is to be able to scan a larger area of the specimen and to be able to image bulk materials. A variety of analytical modes is available for determining the composition and properties of the specimen. The resolution of SEM can range from less than 1 to 20 nm. Hitachi S5500 offers the highest SEM resolution world over and is 0.4 nm at 30 kV and 1.6 nm at 1 kV. User-friendliness is higher for SEM images compared with TEM images.

A typical SEM consists of an electron gun that is capable of emitting electrons thermionically. The electron gun is fitted with a tungsten filament cathode. The electron beam thus generated possesses energy of 40 keV to a few hundred electron volts. It is focused by condenser lenses to a spot of approximately 0.4 to 5 nm in diameter. The scan is made in a raster fashion over the area of the sample surface. The electron beam is allowed to interact with the sample within an interaction volume of 100 nm to 5 μm. The loss of energy from the electrons is by repeated random scattering and absorption within the interaction volume that is teardrop shaped. Elastic scattering, inelastic scattering, and backscattering of electrons take place.

The SEM images the sample surface by raster scanning using electrons. The surface topography, composition, and other properties can be obtained from the scan. Samples must be of a size that can fit into the specimen chamber. The specimen must be electrically conductive. Little sample preparation is needed. Iridescent biological samples are sputter coated with gold for preparation of SEM imaging. The specimen needs to be ground and polished to an ultrasmooth surface in order to use backscattering and quantitative x-ray analysis.

An SEM image of flow-aligned nanowires immobilized on a substrate is shown in Fig. 9.3. Nanowires can be seen to be oriented longitudinally in a single direction along the direction flow during the deposition process. Semiconductor nanowires have gained a lot of interest because of their interesting and novel electrical, chemical, and optical properties. They are used in nanolasers, photovoltaics, and sensor applications such as nano-Chem-FETS. Positioning of these materials has been a technical challenge.

Nanosys patented a method for preparing oriented nanostructures.[3] Nanowires are deposited on a surface substantially in a desired orientation. A fluid containing nanowires is made to flow over a surface. The nanowires

Figure 9.3 SEM image of nanowires (distance captured is 300 μm).

are then immobilized onto the surface with the longitudinal dimensions of the nanowires oriented in the first direction (Fig. 9.3).

9.5 Scanning Probe Microscope

The development of SPM is part of a revolution in the characterization and analysis of materials at the nanoscale. These instruments are different from the optical and electron microscopes. Neither light nor electrons are used to form the image. A topographical map is generated on the atomic scale. The surface features and characteristics of the specimen being studied are obtained. Magnification of more than 1 billion times is possible. Nanoscale features can be examined using SPM techniques. Better resolution is possible with these methods. Three-dimensional images that are amplified are obtained, which contain topographical features of interest. These instruments may be operated in vacuum, air, liquid, or any other desired environment.

A tiny probe with a sharp tip is used and brought in close proximity to within 1 nm of the specimen surface. Across the plane of the surface, the probe tip is then raster scanned. Deflections perpendicular to the plane are experienced in response to electronic interactions between the probe and the surface. The probe movements in and out of the surface are controlled by piezoelectric ceramic components with nanometer resolutions. The motion of the probe tip is also recorded electronically and displayed in the computer monitor using suitable data acquisition techniques. A three-dimensional surface image is then generated. Vacancy defects can be captured using SPM. An atom missing from otherwise regular lattice can be imaged using the SPM technique. Biomolecules and silicon microprocessors have been examined successfully using SPM. It aids in the engineering of nanomaterials.

SPM can be used to prepare nanostructures. DPN uses the SPM tiny probe tip. DPN techniques have been known since ancient times. It is approximately 4000 years old. Ink on a sharp object is transported to a paper substrate by capillary forces. DPN can be used to write patterns consisting of only a collection of molecules. An elastomer stamp may be used to deposit patterns of thiol-functionalized molecules directly onto gold substrates. In this technique, molecules are delivered to a substrate of interest in a positive printing mode. The solid substrate is used as the "paper" and the SPM tiny probe tip is used as the "pen." The probe tip is coated with a patterning compound, the "ink." The desired pattern is produced by application of the patterning compound to the substrate. Capillary transport is the mechanism of delivery of the molecules of the patterning compound to the substrate. A variety of nanoscale devices can be produced using this technique. Software can be developed to have computer-controlled performance of DPN. Many SPM tips are available in the open market. Park Scientific, Digital Instruments, Molecular Imaging, Nanonics Ltd., and Topometrix are some examples of vendors of SPM tiny probe tips. As an aliter, SPM tiny probe tips may be tailor-made to suit the needs of the given application. They can be prepared using e-beam lithography. A solid tip with a hole bored in it can be made using e-beam lithography. SPM tips can be used as AFM tips. The patterning compound is attached to the substrate by physisorption. It can be removed from the surface using a suitable solvent. The physisorption may be enhanced by coating the tip with an adhesion layer and by judicious choice of solvent. The adhesion layer is less than 10 nm in thinness. This layer should not change the tip's surface. The strength can withstand AFM operation of up to 10 nN. Examples of materials that make good adhesion layers are titanium and chromium. Northwestern University[4] has patented a method to use SPM tiny probe tip in DPN to prepare nanostructures.

The SPM tiny probe tip was coated with 1-octadecanethiol. A multistaged etching procedure was described to deposit alkylthiols onto a gold/titanium/silicon substrate. Alkylthhiols form well-ordered monolayers on gold thin films that protect the gold underneath from dissolution during certain wet chemical etching procedures. This is true for DPN-generated resists. The gold, titanium, and silicon dioxide that were not protected by the monolayer could be removed by chemical etchants in a staged procedure (Fig. 9.4). This procedure yields a first-stage three-dimensional feature, multilayered gold topped features on the silicon substrate. Second-stage features were affected by using the leftover gold as an etching resist to allow

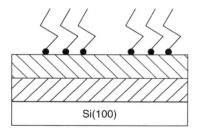

Figure 9.4 DPN deposition and multistage etching procedure to prepare three-dimensional architectures in Au/Ti/Si substrates.

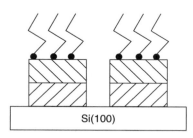

for selective etching of the exposed Si substrate. The remaining gold was then removed to yield all silicon features in the final stage. DPN can thus be combined with wet chemical etching to obtain three-dimensional features on Si(100) wafers with at least one dimension on the sub-100 nm scales. Nanoscale features can thus be prepared on Si wafers. It starts with the coating of 5-nm titanium onto polished single crystalline Si(100) wafers. This is followed by 10 nm of gold by thermal evaporation.

In addition to surface topography data, many other data sets can be obtained using SPMS such as:

1. Using STMs, the conductance and current/distance measures can be obtained.

2. AFM is used for measurement of lateral force and adhesion.

3. Near-field optical microscopy (NFOM) is used for laser transmission at various wavelengths and polarizations.

4. Magnetic force microscopy (MFM) is used for measurement of temperature and parameters.

At University of Tokyo, a direct coupling system between the nanometer and human scale worlds was created. A stereo SEM is augmented with two-handed force feedback control. The sample is positioned within the SEM by the left-hand position and the probe is sampled with a tool by the right hand. A magic wrist was connected to the STM at IBM. The STM translated across the surface the user movements in x and y. The device was kept above the surface. The surface of the gold could be felt by using this technique. System vibrations were noted as a technical problem.

9.6 Microwave Spectroscopy

Phenomena such as Coulomb blockade and single electron tunneling were detected when transport was allowed in small electronic islands in semiconductor nanostructures. The dimensions of these devices are 500 nm^2. They contain 10 to 100 electrons and are referred to as QDs. Transport through these devices was found to be an oscillating pattern of the conductance through the dot as a function of gate voltage. QDs can be used as photon detectors. Measurements of photocurrent induced through single and double QDs have been reported. The millimeter wave radiation in the range of 30 to 200 GHz was coupled to different antenna types. Photon-assisted tunneling through the QDs can be employed for spectroscopy. Rabi oscillations were found to occur when two dots were strongly coupled. When the electromagnetic field interacts with the oscillating valence electron, the oscillations can be probed directly in time-dependent measurements.

The influence of high frequency microwave radiation on single electron tunneling through a single QD was investigated.[5] Tunneling of electrons is allowed. Coupling between radiations to the QD is affected by an on-chip integrated broadband antenna. An additional resonance was attributed to photon-assisted tunneling. Detection principle is based upon absorption of photons by electrons in the leads leading to photon assisted tunneling (PAT). Energy acquired by electrons allows for tunneling above the Fermi energy state. The mechanism of tunneling between two superconductors separated by an insulator is used for high frequency detection. Application of high frequency radiation with frequencies greater than 100 GHz results in modification of the Coulomb blockade oscillations. Electron spin resonance (ESR) in a single QD at filling factor $v \leq 2$ Lamb shift was observed in hydrogen. The effect of electron spin on electron transport through a QD is considered. The QD structure is then imaged using SEM.

PAT can be used as a tool for millimeter wave spectroscopy. Measurements using microwave radiation lend credence to the effect of Coulomb blockade. Coulomb blockade, however, can be overcome with the continuous wave radiation. Time-dependent measurements can be used to obtain details about electronic modes in QDs. In QD systems studied, spectroscopy has been applied in the range of frequency from few MHz to 200 GHz. Recent work reported in the literature demonstrated the ability of integrated broadband antennas to couple continuous wave (cw) millimeter and submillimeter radiation to nanostructures such as QDs and double QDs. The expected PAT resonances in transconductance were shown. The energy acquired by the electrons through absorption of the photons allows for tunneling through states of higher energy above the Fermi energy.

Thus, millimeter wave spectroscopy can be used on quantum point contacts and single and coupled QDs. The influence of potential asymmetries in such devices has been discussed in the literature.

Photon-assisted transport through single and double QDs in the high frequency regime were demonstrated. This method can be applied for spectroscopy of QDs far from equilibrium. This was demonstrated using ESR for a single dot at high magnetic fields. A millimeter wave interferometer was used to perform coherent spectroscopy on coupled QDs. The whole millimeter wave regime is covered and allows for both magnitude and phase sensitive detection. The high frequency conductance through the double QD was detected using the measurements shown. A broadband response was found. Strong vibrations in the relative phase signals were observed when the frequency of radiation was larger than the Rabi frequency of the "artificial molecule." The coherent mode in the double QD is destroyed.

9.7   Auger Electron Microscopy

A scan using an auger electron microscope (AEM) identifies the elemental composition of the analyzed surface. Multiplex scan quantitates the atomic concentration of the elements identified in the survey scans. Detection limits are 1/10 of 1 percent of the atomic composition of the elements. AEM mapping is used for measurement of lateral distribution of elements on the surface. The spatial resolution is approximately 300 nm. Depth profile measurements of distribution of elements as a function of depth into the sample can be obtained. The depth resolution depends on the sample and sputter parameters. Less than 10 nm resolution is possible at a typical sputter rate of 3 nm/min. The sample must be conductive and the sample prepared appropriately. The sample must also be compatible with a high vacuum environment. Typical analysis time is 30 min per sample. Reproducibility is within 10 percent of relative error. Capabilities include measurement of elemental composition of small conductive areas, particles, inclusions, and semiconductor devices.

Auger spectroscopy may be used for chemical specification and chemical bonding characteristics of the material and any adsorbate present before or after exposure to blood in the case of biomaterials used as artificial organs. Biomaterials are synthetic materials used to replace a part of the human anatomy. These have to function in contact with living tissue. Cardiovascular diseases cause a major portion of deaths in the nation. Continuous improvements in the development of biomaterials capable of substituting for parts of the cardiovascular system are important. Examples where implants are used are: heart valves, prostheses, stents, and vascular grafts. Some of the requirements for materials used to prepare these implants are biocompatibility, thrombresistivity, nontoxicity, and durability. A key technical hurdle in interfacing biomaterial with blood revolves around the characteristics of the implant surface. Thrombus and embolism formation at the blood implant interface is a primary concern. Ceramic implants provide chemical inertness, hardness, and wear resistance. They lack the ability to deform plastically.

A multilayered protective coating was formed of ceramic materials. The thickness of the coating layer was in the range of 1 to 100 nm.[6] The coating is comprised of an inner component with one or two layers of zirconia, titania, or alumina. It also contained an outer component formed using a water swellable ceramic material capable of forming a hydrate or hydroxide compound upon contact with an oxygen containing environment. The outer component is made of aluminium, zirconium, or halfnium. The overall thickness of the coating is several microns.

9.8 Raman Microscopy

The Raman microscope is designed based upon the principle of the Raman effect. Raman was given the Nobel Prize for physics in 1930 for his work on molecular scattering of light. He explained why the sky and seas were blue in color. He showed that molecules scatter light. The scattering depends on the characteristics of the molecule. Raman microscopes have been invented to back out the molecular type from the scattering information. Molecular scattering of light was primarily from molecular disarray in the medium and due to the local fluctuations of optical density. The effect could be seen in vapor and gases and in crystalline and amorphous solids.

The Raman effect occurs when a molecule is excited by light. The light interacts with the electron cloud of the bonds of that molecule. The incident photon excites one of the electrons into the virtual state. The molecule is excited from the ground state to a virtual state and relaxes into a vibrational excited state. This generates Stokes Raman scattering. The scattering provides a fingerprint from which a molecule can be identified. The changes in chemical bonding can be studied such as when a substrate is added to an enzyme. Raman spectroscopy can be used to measure temperature and find the crystallographic orientation of a sample.

Intel has patented a microfluidic apparatus and methods for performing molecular reactions with nucleic acid molecules and Raman microscopic systems to detect these molecules.[7] The determination of the human genome sequence in its entirety has lead to progress in identifying the genetic basis for diseases such as cancer, cystic fibrosis, sickle cell anemia, and muscular dystrophy. Current methods for measuring nucleic acid sequence information are tedious and expensive and often times inefficient. Lack of harmonization in multimolecule polymer reactions such as in the case of exonuclease sequencing of DNA results in imprecise results. Intel developed an improved apparatus and method for performance of biomolecular reactions. It allows for the capture of a single nucleic acid molecule in a microfluidic channel upstream from an optical detector. Sequential detection of one or more nucleotides is made using a Raman microscope. Optical tweezers are used in isolation of a single nucleic acid molecule. Lasers that are used by optical tweezers may interfere with the detection using a Raman microscope. Integration of optical tweezers and Raman microscope may limit the field of view.

The Intel apparatus does not allow for interference of Raman microscope detection capabilities. Solid supports such as particle or bead can be used to allow for immobilization of a single nucleic acid molecule on them. Optical tweezers are usually a gradient force optical trap such as a single beam gradient force optical trap that captures the single particle downstream from the laser beam. Single nucleotides can be cleaved from the bead using an exonuclease. Single nucleotide molecules are detected using SERS. The inclusion of a restriction barrier in a microfluidic channel and the immobilization of an optically transported bead allow for removal of the optical tweezers from the optical path of the detection device. Thus, the interference from the additional light source of the optical tweezers close to the collection volume of the detector is obviated.

The SERS detection unit includes a detection light source, typically a laser light source used for irradiation of the molecule, and a detection unit for capturing Raman emission from the emitting molecule. Thus, the nucleic acid molecule attached to the surface of the particle is analyzed. The system can be used to obtain the sequence distribution of the nucleic acid molecule. The movement of the nucleic acid molecule is restrained. The molecule is then contacted with an agent that removes nucleotides such as an exonuclease. A terminal nucleotide is released and then detected using SERS. Raman emission from the first and second released nucleotides is detected. PCR can be analyzed in a similar fashion using hybridization reactions, fluorescent probes, and Raman labeled molecular probes.

The apparatus and system (Fig. 9.5) consists of: (1) a light source, (2) a detector to detect SERS emission of a molecule excited by the

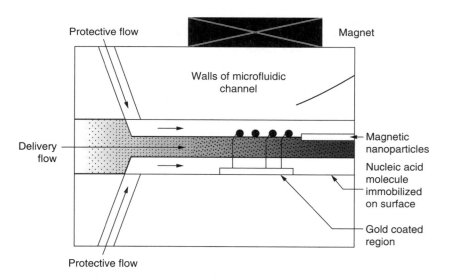

FIGURE 9.5 Microfluidic system to perform nucleic acid molecular reactions.

light source, and (3) a first channel that provides a restriction barrier. The length of the nucleic acid molecule can vary from 10 base pairs to 5 million base pairs.

9.9   Atomic Force Microscopy

STM evolved into the AFM. AFM is a special type of SPM, discussed in Sec. 9.5. The resolution achieved using AFM is a fraction of a nanometer.

The AFM was invented in 1986 and is used in nanoscale characterizations. A schematic of a typical AFM assembly is shown in Fig. 9.6. The specimen surface is scanned using a microscale cantilever with a probe at its end with a sharp tip. The tip radius of curvature is of the order of a few nanometers. The forces between the tip and the sample surface cause the deflection of the cantilever. The deflection is found to obey Hooke's law of elasticity. Different types of forces can be measured using the AFM such as van der Waals, hydrogen bonding, surface tension, magnetic, and electrostatic. The laser source excites the specimen surface. The cantilever reflects the laser light and is captured by the array of photodiodes. Cantilevers are made up of piezoresistive elements such as a strain gauge and a Wheatstone circuit is used to measure the deflection. The tip to sample distance is controlled using a feedback control loop. AFM can be operated in the imaging or tapping mode. Individual atoms can be imaged using AFM. Atoms of silicon, tin, and lead on an alloy surface can be distinguished using AFM.

AFM has been used to form complex patterns in thin layers of MoO_3 grown on the surface of MoS_2.[8] The pattern lines formed by

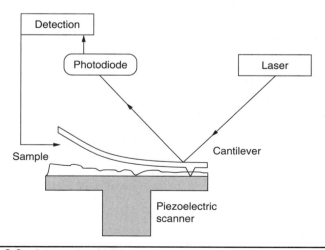

FIGURE 9.6   Schematic of AFM.

using AFM were found to be less than 10 nm. The resulting structure is imaged using the AFM as well. AFM has been used as a tool to move atoms or clusters of atoms directly into a desired configuration. Orientation ordering of organics can be affected by direct contact imaging of soft organic layers under sufficiently high loads. The scale of modifications is greater than 100 nm. The limits of direct surface manipulation were explored using the AFM. The system they used was a thin 50 A metal oxide film made up of MoO_3 on the surface of MoS_2. Harvard University's patent has some advantages over previous methods such as: (1) rigid and nondeformable thin MoO_3 film compared with organic layers, (2) MoO_3 can be machined or imaged depending on the applied load of the AFM cantilever, and (3) MoS_2 substrate acts as a stop layer. Patterns with less than 10 nm resolution can be achieved in this manner. Crystallites of α-MoO_3 were grown on the surface of single crystal 2H-MoS_2 by thermal oxidation using purified oxygen at 480°C for 5 to 10 min. AFM images along with TEM and x-ray photoemission spectroscopy were used to characterize the nanostructures. The crystallites grow with the perpendicular axis to the substrate layer. Crystallites 1 to 3 unit cells thick and 200 to 500 nm on edge are formed.

The world's highest resolution AFM as of December 2008 was developed by Asylum Research in the name of Cypher. Cypher is designed as a closed-loop system in all three dimensions in order to achieve atomic resolution. Features include automatic laser alignment, interchangeable light source modules with laser spot sizes as small as 3 µm, and cantilevers smaller than 10 µm.

9.10 Helium Ion Microscopy

The HeIM was developed as an alternative to the electron microscope. Helium ions are used in place of electrons for imaging. Helium ions possess shorter wavelengths than electrons. They can form more tightly focused beams. This translates into better image resolution. Orion is the first commercially available HeIM and it was installed in the National Institute of Standards & Technology from Carl Zeiss, Peabody, Massachusetts. Carl Zeiss has broken the world record for resolution. TEM-like resolution on bulk samples with SEM ease can be achieved using the Orion plus microscope. A surface resolution of 0.24 nm has been achieved reproducibly using Orion HeIM. The resolution is close to the diameter of a single atom and 3 times better than the SEMs with the same surface sensitivity. A new benchmark was achieved for surface imaging in the subnanometer range. Semiconductor manufacturers who need to see small features that currently cannot be resolved use the HeIM. Advances in miniaturization of feature sizes of semiconductor devices require high resolution microscopy as a must. Some layers of ICs have reached a thickness of only a few atoms. Images with Rutherford backscattered ions (RBIs) can be created using HeIM. RBIs are high energy helium ions that rebound off a sample. Information on chemical composition can be obtained using HeIM that cannot be obtained using SEM. The chemical composition of a defect in a semiconductor chip can be detected from HeIM.

The source of the microscope is small and the helium ions emanate from a region as small as a single atom. Because of the lower wavelength, the helium ions do not suffer appreciably from adverse diffraction effects. Adverse diffraction effects are a law of physics that predicts a fundamental limit on the imaging resolution of electrons. Signals are triggered directly from the surface of the sample by the helium ion beam. They stay collimated upon entering the sample. As a result, sharp and surface sensitive images at the quoted resolution can be obtained. In a SEM, a majority of secondary electrons that are used for imaging stem from deeper, less confined regions within the sample causing blurrier images with less resolution compared with images from HeIM. The cost of a HeIM in 2008 was $2 million.

HeIM is related to field-ion or field-emission microscopy. These were used for looking at individual atoms on a cryogenically cooled tungsten tip in an ultrahigh vacuum system with small amounts of helium gas present in it. A gallium ion microscope can be used to analyze a wider range of samples. The problem of sputtering of the sample prior to imaging remains with HeIM. A distinct pyramid-shaped ion source allows helium ions to form a more tightly focused beam leading to images with better resolution. Helium ions are lighter than gallium ions and the samples are not allowed to deteriorate readily.[9] Helium as an ion source is the primary component of HeIM. Images can be collected in two modes: the RBI and secondary electron mode. A combination of heat and electric fields is used to allow for emission of electrons. The signals are synthesized into an image at the detector. Helium ions hit the sample. Upon a hit, secondary electrons and RBIs are released. Helium ions possess greater mass and much shorter wavelength compared with electron beams. Helium ions interact more strongly with the materials compared with electrons and 100 times more secondary electrons are produced. More information goes to the detector resulting in images with greater detail. Researchers in nanoscale materials are eager to use HeIM to obtain information on chemical composition from RBI images. Useful information about sample surfaces can be obtained using secondary electron images.

Automated microelectronic chip manufacturing can use HeIM for monitoring in place of SEM. One weakness of HeIM is that more damage to the specimen is caused in HeIM compared with electron microscopes. Properties of cellulosic nanocrystals and CNTs can be examined using HeIM. Chemistry of nanowires and other nanoscale materials was examined using HeIM at Harvard University, using both the RBI and secondary electron modes.

Inspection of nanofabricated structures and nanomanufacturing can be affected using HeIM. Monolayers of graphene and images of dry biological samples can be studied using HeIM.

9.11   Summary

Needs for characterization of nanostructures are on the rise. Resolution limits of optical microscopes are of the order of a wavelength of light. Per the Raleigh criterion, the resolution limit of optical microscopes is of the order of 200 nm. In order to characterize nanoscale materials, x-ray and helium ion microscopes are needed. Optical microscopes, SEMs, TEMs, SPMs, and HeIMs have increasing powers of resolution in the order mentioned.

Structural information in the scale of 2 to 25 nm can be characterized using SAXS. Monochromatic sources of x-rays are used to excite the sample and scattered x-rays are detected by a two-dimensional flat x-ray detector. Structure is deduced from patterns in the scatter. Interpretation of scattered pattern can be accomplished using Porod's law, Guinier approximation, Fourier transformation, etc. Thin films, multilayered systems, oriented nanoparticles with different chemical compositions, colloids, protein solutions, nanocomposites, micelles, and fiber structures can be studied using SAXS. WAXS, GISAXS, and SWAXS are techniques that are variations of SAXS.

TEMs have higher resolution power. Sample preparation for TEM analysis is complex and the thickness of the sample has to be down to a few hundred nanometers. An electron beam is produced from a tungsten filament subjected to a high voltage. Electrons are allowed to pass through the specimen. With HRTEM, resolutions achievable are as small as 1 A. TEM is used in life sciences, biomedical investigations, as a diagnostic tool in pathology, and for the imaging of atoms, oligopeptides, nanogold, and self-assembled nanotubes. It can also be used as an elemental analysis tool in addition to EDXA and at low temperatures as cryo-TEM.

Magnification in SEM ranges from 25 to 250,000 and resolution size is down to 1 to 25 nm. An electron gun is used to generate electron beams. Spatial resolution of SEM depends on the wavelength of the electron and an electro-optical system that produces the scanning beam. Resolution at an atomic scale is not possible as can be with the case of TEM and HeIM. The electrons generated are focused as a spot with nanoscale dimensions. Upon impingement with the specimen, electrons undergo elastic scattering, inelastic scattering, and backscattering. Raster scanning is used to image surfaces. Surface topography, composition, and other properties can be obtained from a raster scan.

A topographical map on an atomic scale can be generated using SPM. Neither electrons nor light is used for formation of images. Magnification of higher than 1 billion times is possible. A tiny probe with a sharp tip is brought in close proximity to within 1 nm of a specimen surface and then raster scanned. Nanoscale defects, biomolecules, and silicon microprocessors can be characterized using SPM. They can also be used to prepare nanostructures. DPN is such a technique. A tiny probe tip is used as "pen" to write structures consisting of a few molecules. STMs can be used to obtain conductance and current/distance measurements, AFM can be used for lateral force and adhesion measurements, NFOM can be used for laser transmission at various wavelengths, and MFM can be used for temperature and other parameters.

QDs contain 10 to 100 electrons in devices of dimensions of 500 nm². Photocurrent induced in QDs can be measured. The influence of high frequency microwave radiation on single electron tunneling through a single QD was used in microwave spectroscopy. Elemental composition of test specimens can be obtained using an AEM. Detection limits are 0.1 percent of the atomic composition of the elements. Spatial resolution is approximately 300 nm. Depth resolution is approximately 10 nm and typical analysis is 30 min per sample.

Raman microscope is designed based on the Raman effect. Molecular types can be obtained from the scattering information. Raman microscopes can be used to study gene expression and DNA sequence distribution. SERS can be used to detect single nucleotide molecules. SERS includes a laser light source that excites the molecule and a detection unit for capturing Raman emission emanating from the molecule.

STM evolved into AFM (Fig. 9.6). Fraction of nanometer resolution can be achieved using AFM. A specimen surface is scanned using a microscale cantilever with a probe at its end with a sharp tip with a radius of curvature of a few nanometers. A laser source excites the sample. The deflected cantilever reflects the laser light. The reflected light is captured by an avalanche of photodiodes. Individual atoms can be imaged using AFM.

HeIM was developed as an alternative to the electron microscope. Helium ions possess shorter wavelengths compared with electrons. Orion HeIM achieved a space resolution of 0.24 nm. This is close to the diameter of a single atom and 3 times better in resolution compared with electron microscopes. Individual atoms can be looked at. HeIMs can be operated in RBI and secondary electron mode.

Review Questions

1. Sketch a wave of light at a wavelength of 400 nm. Show a nanoparticle along the path of the optical wave. What would you expect to see as the particle's image?

2. What are the resolution limits achievable using optical microscopes according to the Raleigh criterion?

3. Rank the following microscopes according to their powers of resolution: (a) SEM, (b) SPM, (c) TEM, (d) optical microscopes, and (e) HeIM.

4. Explain the importance of scattering pattern in SAXS.

5. What happens to the noise in the pattern as the desired angle is decreased during SAXS measurements?

6. What is Porod's law?

7. Explain what is meant by Guinier approximation.

8. Why are Fourier transforms of the form factor of SAXS measurements obtained?

9. How is the structure factor used in identification of short-range order?

10. Distinguish between GISAXS and WAXS.

11. Can phase state of materials be determined using SAXS?

12. During the analysis using SAXS, what happens to the sample both physically and chemically?

13. Can self-assembly of molecules and micelles formation be studied using SAXS?

14. Can nanostructures be prepared using SAXS?

15. What are the components used in SAXS?

16. Compare the specimen preparation needed in TEM and SEM.

17. What is the role of electrons in TEM?

18. Distinguish between HRTEM and STEM.

19. What is meant by raster scanning?

20. What is accomplished during EDXA and cryo-TEM procedures?

21. What is the ideal characterization tool to study branched supramolecular structures?

22. What is the ideal characterization tool for DNA and protein structures?

23. What is an ideal characterization tool for study of reverse structures?

24. What is an ideal characterization tool for study of oligopeptides?

25. What is an ideal characterization tool for study of reverse micelle formation?

26. What is the role of electron channeling contrast in the design of SEM?

27. Compare the electron gun used in SEM with the tungsten filament charged with a high voltage source in TEM to generate electrons.

28. Is image magnification a function of the objective lens in SEM analysis?

29. Why is resolution of atomic dimension not possible using SEM as compared with TEM?

30. Which instrument is more user-friendly: SEM or TEM?

31. Why are SEMs preferred to study nanowires?

32. Why is cryo-TEM preferred to study self-assembled structures?

33. How is the design of SPM different from SEM in concept?

34. What is the role of deflections in SPM analysis?

35. Elaborate on DPN to create nanostructures.

36. How is DPN different from e-beam lithography?

37. What kind of chemical reactions take place during creation of nanostructures using DPN?

38. What kinds of datasets can be obtained using: (a) STM, (b) AFM, (c) NFOM, and (d) MFM?

39. What is done differently in the technique called stereo SEM?

40. How is microwave spectroscopy better suited for characterization of QDs?

41. What is the role of PAT in operation of millimeter wave spectroscopy?

42. How is the elemental composition of surface obtained using AEM different from that obtained using SPM?

43. What is meant by depth profile measurements?

44. What is meant by depth resolution and why is this an important consideration in AEM analysis?

45. How can the chemical bonding characteristics of materials be better-studied using AEM analysis?

46. What is meant by the Raman effect?

47. How is the "fingerprint" of molecules detected using a Raman microscope?

48. Can temperature measurements be made using Raman microscopes?

49. Discuss the Intel patent that uses a microfluidic apparatus to detect molecular reactions with nucleic acid molecules.

50. What is the significance of using Raman microscopes in the study of biomolecular reactions such as PCR?

51. How do the cantilever deflections in AFM analysis affect the passage of laser beams from excitation source to the specimen to the detector?

52. How is AFM used in the creation of nanostructures?

53. Why are better resolutions achievable using HeIMs?

References

1. P. Dubcek, Nanostructure as seen by the SAXS, *Vacuum*, 80, 92–97, 2005.
2. S. Zhang and S. Vauthey, Surfactant Peptide Nanostructures and Uses Thereof, US Patent 7,179,784, 2007, Massachusetts Institute of Technology, Cambridge, MA.
3. X. Duan, R. H. Daniels, C. Niu, V. Sahi, J. M. Hamilton, and L. T. Romano, Methods of Positioning and/or Orienting Nanostructures, US Patent 7,422,980, 2008, Nanosys Inc., Palo Alto, CA.
4. C. A. Mirkin, R. Piner, and S. Hong, Methods Utilizing Scanning Probe Microscope Tips and Products Therefore or Produced Thereby, US Patent 6,827, 979, 2004, Northwestern University, Evanston, IL.
5. R. H. Blick, Microwave spectroscopy on quantum dots, in *Handbook of Nanostructured Materials and Nanotechnology*, H. S. Nalwa, Ed., Academic Press, Elsevier Science, Amsterdam, Netherlands 2000, pp. 309–343.
6. C. R. Aita, V. V. Yakovlev, M. M. Cayton, M. Mirhoseini, and M. Aita, Self Repairing Ceramic Coatings, US Patent 6,869,701, 2005, C. Aita, Shorewood, WI.
7. N. Sundarajan, L. Sun, Y. Zhang, X. Su, S. Selena Chan, T. W. Koo, and A. A. Berlin, Microfluidic Apparatus, Raman Spectroscopy Systems, and Methods for Performing Molecular Reactions, US Patent 7,442, 339, 2008, Intel Corp., Santa Clara, CA.
8. C. M. Liebler and Y. Kim, Machining Oxide Thin-Films with an Atomic Force Microscope: Pattern and Object Formation on the Nanometer Scale, US Patent 5,252, 835, Harvard University, Cambridge, MA.
9. R. Petkewich, Say hello to helium ion microscopy, *Chemical and Engineering News, Science and Technology*, 86, 47, 38–39, 2008.

Index